In the Beginning

Also by Gerard M. Verschuuren
from Sophia Institute Press:

Forty Anti-Catholic Lies
A Mythbusting Apologist Sets the Record Straight

Dr. Gerard M. Verschuuren

In the Beginning

A Catholic Scientist Explains
How God Made Earth Our Home

SOPHIA INSTITUTE PRESS
Manchester, New Hampshire

Sophia Institute Press
Box 5284, Manchester, NH 03108
1-800-888-9344

www.SophiaInstitute.com

Sophia Institute Press® is a registered trademark of Sophia Institute.

Library of Congress Cataloging-in-Publication Data
To come.

First printing

To Stephen M. Barr,
an unwavering defender
of modern science
and ancient faith

Contents

Introduction ix

Part 1

Before We Get Started

1. The Timescales 3

2. Dating Techniques 11

Part 2

Where Science and Religion Meet

3. Creation or the Big Bang? 19

4. What Comes with Creation? 39

5. The Evolution of the Universe 65

6. How the Earth Developed 79

7. The Evolution of Life 91

8. Were We Meant to Be Here? 133

   A Final Word 153

   Appendix: Is Planet Earth Unique? 167

   Index 177

   About the Author 181

Introduction

When, on July 20, 1969, a spacecraft landed on a rocky basalt crater of the moon, one of its two astronauts would recall, "It suddenly struck me that that tiny pea, pretty and blue, was the Earth. I put up my thumb and shut one eye, and my thumb blotted out the planet." It is this tiny pea in a vast universe that we call our home.

How did the earth become our home? It certainly didn't happen overnight. We all know that Rome was not built in a day. Well, neither was the earth, a tiny speck in an immense universe. It took billions of years to get our home set up in the proper way so we could enter and comfortably live here. Some see this as the outcome of a protracted sequence of coincidences. Others think this process had to be the way it was, because it was intended to be that way. Who is right?

Indeed, on the surface, it seems to be a lengthy process of trial and error with numerous steps of a seemingly hit-or-miss character. Yet the entire process was much more streamlined than it appears. How could that be? In this book, we will discover that there was, and still is, quite an unmistakable pattern that has been steering the entire process. There is reason to believe this pattern came from God, so the entire process was and is under God's control. That's how God made a home for us in this immense universe.

In the Beginning

This book is not written for specialists. Its subject requires some scientific terminology and some basic explanations of scientific issues, and I added notes so that you can check certain claims I make, but I wanted to keep the text as readable as possible for the average person. I hope I succeeded.

Part 1

Before We Get Started

When you want to delve into the history of the universe, of the earth, and of life on it, you will soon discover there is quite some disagreement about the factual data. First, there is disagreement about the time scales of the events involved, ranging from very short to very long. Second, there is not a commonly accepted standard for determining when those events occurred. Those two issues need to be settled before the real discussion can begin.

1

The Timescales

The Book of Scripture

A few centuries ago, James Ussher (1581–1656), the Anglican archbishop of Armagh in the Church of Ireland, calculated a chronology of the history of the world based on data he had gathered from Scripture. Based on a literal reading of the Old Testament, Ussher deduced that the first day of creation began at nightfall on Saturday, October 22, 4004 BC.[1] If Ussher could be that exact, many think, then his calculation must have been very reliable and accurate.

What Ussher didn't realize, however, is that the Book of Scripture is not a history book or a science book. It's not a book about the history of the universe and the history of the earth. It does not tell us how old the universe is or how old the earth is. Ussher made the same mistake that many fundamentalist Protestants make: he read Scripture as if it were a science textbook.

In contrast, the Catholic Church has made, for many centuries, an important distinction between what she calls the Book of Scripture and the Book of Nature. In making this distinction, the Church follows a long tradition. This distinction can be traced back

---

[1] James Ussher, *Annals of the World*, 1.

at least to St. Augustine, who said around the year 400: "It is the divine page that you must listen to; it is the book of the universe that you must observe."[2]

The Catholic Church acknowledges that God speaks to us in two ways: through the Book of Scripture and through the Book of Nature—and He is the source of both. Galileo used the metaphor of the two books to defend the compatibility of his new astronomy with Sacred Scripture. Galileo presented Nature and Scripture as two books proceeding from the same Divine Word. Therefore, the glory of God can also be known by means of the works that He has written in what Galileo called the "open book of heaven." The astronomer Johannes Kepler, a contemporary of Galileo, also spoke of the Book of Nature as a book in which God reveals Himself in another way from the way He is revealed in Scripture. In his own words, "I believe that together with the Holy Scriptures came the book of Nature."[3] And very recently, Pope Benedict XVI told us to see "nature as a book whose author is God in the same way that Scripture has God as its author."[4]

That's why there can be a big difference between what the Book of Scripture tells us and what the Book of Nature tells us. To use the metaphor Cardinal Baronius introduced at the time of Galileo, "The Bible teaches us how to go to heaven, not how the heavens go."[5] Put differently, the Book of Nature tells us how we

---

[2] Augustine, *Enarrationes in Psalmos*, 45, 7.

[3] Johannes Kepler, *Epitome of Copernican Astronomy*, vol. 16 of Great Books of the Western World (Chicago: Encyclopedia Britannica, 1952), 850.

[4] Benedict XVI, Address to the Pontifical Academy of Science (October 31, 2008).

[5] This remark, which Baronius probably made in conversation with Galileo, was cited in Galileo's 1615 *Letter to the Grand Duchess Christina*.

are rooted in nature, whereas the Book of Scripture deals with how we are rooted in God.

In spite of this important, long-standing difference, James Ussher read the Book of Scripture as if it were the Book of Nature. He did not heed St. Augustine's warning that it is "dangerous to have an infidel hear a Christian ... talking nonsense."[6] To avoid talking nonsense, we must also pay attention to what the Book of Nature tells us. St. Augustine could not have said it more clearly: "It not infrequently happens that something about the Earth, about the sky, ... about the nature of animals, of fruits, of stones, and of other such things, may be known with the greatest certainty by reasoning or by experience, even by one who is not a Christian."[7] That's why scientists—even if they are not Christian—can tell us much about the age of the universe and the age of the earth.

The Book of Scripture, on the other hand, has an entirely different approach. The creation account in the book of Genesis, chapter 1, is not about an event that happened long ago. Neither is it an event that lasted six days—not even if we take each of those six days as a much longer period of time. It's hard, if not impossible, to read Genesis 1 as a *chronological* account that treats God's creation as if it were a six-day event.

Creation cannot follow a timeline, for time itself is something God had to create first. Creation creates chronology, but it cannot become part of chronology, nor can chronology be the framework of creation. Considerations like these are troublesome for a chronological interpretation of Genesis 1.

There are other indications that tell us that we are not dealing with a chronological order: the creation of the sun happens three days after the day-night cycle is established. In addition, plants

---

[6] Augustine, *The Literal Meaning of Genesis*, 1, 20.
[7] Ibid., 1:19.

In the Beginning

ı, but they were created one day before the sun was created. Perhaps these are minor problems, but they cannot be ignored.

So, we can draw the provisional conclusion that Genesis 1 is not meant to be understood as a literal *chronological* account. As Pope Pius XII wrote: "The ancient peoples of the East, in order to express their ideas, did not always employ those forms or kinds of speech which we use today; but rather those used by the men of their times and countries. What those exactly were, the commentator cannot determine as it were in advance, but only after a careful examination of the ancient literature of the East."[8]

This leads to the second possibility—that Genesis 1 is to be given not a chronological, but a structural reading. The Catechism of the Catholic Church (CCC) explains that "Scripture presents the work of the Creator symbolically as a succession of six days of divine 'work,' concluded by the 'rest' of the seventh day (Gen. 1:1–2:4)" (337).

The structural approach is not a new, modern idea. For many centuries, it has been recognized that the six days of creation are divided into two sets of three. In the first set, God separates one thing from another: On day one, He separates the light from the darkness (thus giving rise to day and night); on day two, He separates the waters above from the waters below (thus giving rise to the sky and the sea); and on day three, He separates the waters below from each other (thus giving rise to dry land in between the waters). Classically, this section is known as describing the work of division.

In the second set of three days, God goes back over the realms He produced by division during the first three days and populates, or adorns, them. On day four, He adorns the day and the night with

---

[8] Pius XII, Encyclical Letter *Divino Afflante Spiritu* (September 30, 1943), 36.

the sun, the moon, and the stars. On day five, He populates the sky and the sea with birds and fish. And on day six, He populates the land (between the divided waters) with animals and mankind. Classically, this is known as describing the work of adornment.

It is through this structure that Genesis 1 proclaims its core message of monotheism against the pagan beliefs that surrounded Israel. It emphatically proclaims that nothing exists that does not owe its existence to God the Creator. The Book of Scripture teaches us not science but belief in one God, the Maker of Heaven and Earth.

Most of the peoples surrounding Israel regarded the various regions of nature as divine: they worshipped the sun, the moon, and the stars as gods; they had sky gods, earth gods, water gods, and gods of light and darkness, of rivers and vegetation, of animals and fertility. To battle this, Genesis 1 proclaims its radical and imperative affirmation of monotheism. Each day dismisses an additional cluster of deities: On the first day, the gods of light and darkness; on the second day, the gods of sky and sea; on the third day, earth gods and gods of vegetation; on the fourth day, sun, moon, and star gods (including astrology); on the fifth and sixth days, gods from the animal kingdom (such as sacred falcons, lions, serpents, and golden calves). Finally, even humans are emptied of any intrinsic divinity, even while they are granted a divine likeness. So, each "day of creation" shows us another set of idols being smashed. Nothing on Earth is a god, but everything comes from God. Obviously, the issue at stake in Genesis 1 is idolatry, not science; mythology, not natural history; theology, not chronology; theology, not scientology.

Once we know this, Ussher's calculations make no sense at all. He thought he could determine the age of our world and the universe by counting back from the last page in the Bible to its first page. He read the Book of Scripture as if it were the Book of Nature. So, his idea that the first day of creation began at nightfall on Saturday, October 22, 4004 B.C. is an embarrassing misinterpretation.

In the Beginning

The Book of Nature

The Book of Nature tells us a story quite different from what James Ussher came up with. The timescale of the Book of Nature is very different from the timescale used in the Book of Scripture. In the Book of Nature, we need to work with different, scientific dating techniques. What do they tell us?

As we will see in detail later, astrophysics estimates that the Big Bang happened approximately 13.8 billion years ago, which is thus considered the age of the universe. Astronomy tells us the age of the earth is 4.5 billion years, based on evidence from radiometric age dating of meteorite material. Geology adds to this that ancient rocks exceeding 3.5 billion years in age are found on all of the earth's continents. (The earth's oldest rocks have been recycled and destroyed by the process of plate tectonics, which we discuss later.) And paleontology shows us that fossils range enormously in age and can be as old as 4.1 billion years.

The time spans we are dealing with here are so huge that they defy human imagination. We cannot really fathom how immense these numbers are. What may help us to get a better feel for such enormous magnitudes is using the timeline of a twenty-four-hour *day*, with the Big Bang taking place at the beginning of the day, at 12:01 a.m., and the current time at the end of the day, at 11:59 p.m. On this scale, each hour of the day represents some 600 million years. That's still beyond our imagination, but at least it makes it easier for us to compare how long it took for certain events to take place. On this scale of a day, Planet Earth begins to form at 4:17 p.m., and life begins to develop at 5:08 p.m. Seen this way, humanity appears on Earth at the very last minute—not sooner than 11:59 p.m. It is a "last minute" step, some say—just in time, or perhaps almost too late.

Using the timeline of one calendar *year*, instead of a day, would amount to about 38 million years per day. After the Big Bang occurred at the beginning of January, planet Earth would then start to form at the beginning of September. Then life would emerge on Earth during the middle of September, dinosaurs would appear around December 24, and mammals around December 28. Then, even closer to the end of the year, human beings would emerge around December 29. Could you have imagined a more delayed arrival for humanity?

Given such enormous timescales, it seems very reasonable to ask why everything took so long. Many people, especially scientists, believe that such long periods are required for processes that are considered to be entirely coincidental. The more technical term is *random*—a matter of hit-and-miss or trial and error. Their reasoning is based on a statistical law: the "law of large numbers" (LLN).

We are all familiar with that law, perhaps without even knowing it. Given enough time at a slot machine, we think, we may have a chance eventually to hit the jackpot. We also know that we have a better chance of winning the lottery if we buy lots of tickets at once, or only a few but repeatedly for a long period.

Many people apply this law also when it comes to the enormous time span of our universe. They point out that chance and randomness need large numbers and long periods of time to perform their hit-and-miss or trial-and-error actions. So, the odds of a planet on which life is possible and on which life did emerge become better and better when we increase the number of years dramatically. That explains, according to these people, why the earth had to be so old and why the universe had to be much older. It is chance or randomness that many people see as the best candidate for the origin of life, of the earth, and of the universe. It's basically the idea that if you try long enough, you will eventually win the lottery.

Whereas some see the emergence of humanity on Earth as the outcome of a long, protracted sequence of coincidences, however, others think that this process had to be the way it was, simply because it was "intended" this way. The latter view implies that human beings appeared after everything had been meticulously prepared for them—which took a long, long time of detailed preparation.

We will see later who is probably right. For now, I will say that, most likely, randomness did play a role in the evolution of the universe, the earth, and life. Even given the enormous timescales we are dealing with here, however, randomness could probably not do the entire job of making the earth our home. There must be more to it. That's what we investigate in the rest of this book.

2

Dating Techniques

The Universe

How do we know the age of the universe if no one was there at the beginning? There are several ways we have come to an approximate age of 13.8 billion years.[9] One of the measurements used to calculate the approximate age of the universe is based on the expansion rate of the universe, which is the increase in the distance between two distant parts of the universe with time. Scientists can use the Hubble Space Telescope to measure the rate of this expansion. Based on this rate, they can extrapolate backward in time—to the point at which all the mass of the universe was concentrated in a single point, which was the event of the Big Bang. (There are other ways, based on some very technical calculations, but they go far beyond the scope of this book.)

If we know the expansion rate of the universe, we have a better idea of when the universe was born. The first reasonably accurate measurement of the rate of expansion of the universe—a numerical value now known as the Hubble constant—was made

---

[9] Planck Collaboration, "Planck 2015 Results. XIII. Cosmological Parameters," *Astronomy and Astrophysics* 594 (2015): A13.

In the Beginning

in 1958 by astronomer Allan Sandage.[10] His measured value for the Hubble constant came very close to the value range generally accepted today.

At first, calculations made this way faced problems with the data. The first estimate of the age of the universe came from the calculation of when all the objects must have started speeding out from the same point. These first estimates based on the rate of expansion gave about a billion years, which was inconsistent with the known ages of the oldest rocks and of stars. But those first estimates turned out to be based on a mistaken calculation of the distances between galaxies. Science is always a work in progress!

Eventually things got sorted out, and evidence for the age of the universe was built up to the point at which it is regarded as pretty conclusive. Although nothing in science is final, cosmologists tell us that an approximate age of 13.8 billion years for the universe is still the latest and best we have. It remains standing for now until further notice.

That's how we know when the universe was born. But what about the earth? How do we know when the earth was born and when the earth would hold its first fossils as an indication of life?

The Earth

Well, planet Earth and some other planets and moons in the solar system have sufficiently rigid structures to have preserved records of their own histories. But how do we know when those records were preserved? Geologists used to construct a geologic table based on the relative positions of different geological layers in the earth's crust. In general, fossil formation has a better chance of happening

---

[10] A. R. Sandage, "Current Problems in the Extragalactic Distance Scale," *Astrophysical Journal* 127, no. 3 (1958): 513–526.

in water (away from oxygen in the air), as bones are buried by sediment, sand, and other materials that settle out of the water. In time, more deposits cover the layers of mud and sand, thus pressing down the layers holding the remains and hardening them into rock. Later, changes in the earth's surface may expose parts of these layers, as when stream erosion wears away layers of sediment. Unfortunately, sequences of layers often become eroded, distorted, tilted, or even inverted after deposition. No wonder the ages of various rock strata and the age of the earth were subjects of considerable debate.

Not long ago, fossils could be dated only by their appearance in the sequence of geological layers—deeper is older; higher is younger. But because these layers have been shuffled around quite a bit in the history of the earth, often the date of the layers had to be determined by the fossils they contained—which is obviously a case of circular reasoning. This way we can get, at best, a relative way of dating, but no absolute dates. Fortunately, we have much better ways of dating nowadays. The most important technique is radioactive dating. Here follows a very basic explanation of isotopes first.

On Earth we know of ninety-two naturally occurring chemical elements. (Other elements have only been synthesized in laboratories or nuclear reactors.) The periodic table of elements shows all the elements in order, based on their atomic numbers. The number of protons determines the atomic number (1–92) and defines the element and its chemical behavior. For example, carbon atoms have six protons, hydrogen atoms have one, and oxygen atoms have eight.

Every chemical element has its own kind of nucleus, made up of a specific number of protons—its atomic number—together with some number of neutrons, which differs in different "isotopes" of the element. An atom of the element carbon, for instance, has six protons in its nucleus. One isotope of carbon has also six neutrons, making a total of twelve particles. Therefore, it is called "carbon 12." Another carbon isotope has eight neutrons, which makes it "carbon 14" (6 + 8 = 14). The simplest and smallest element is hydrogen, with one proton and one neutron in its nucleus, "hydrogen 2."

How can these elements help us to date things on Earth more accurately than strata or geological layers can do on their own? The secret can be found in the existence of isotopes. As said before, isotopes of a particular element differ in the number of neutrons in their nucleus. Some isotopes are inherently unstable. That is, at some point, an atom of such an isotope will undergo radioactive decay and spontaneously transform into a different isotope.

Certain isotopes have a known and constant rate of radioactive decay. These isotopes always start with a known level of radioactivity when they cool from molten lava and become solid. So, all we have to do is measure the current level of radioactivity and then figure out the amount of time it took to get from the initial level of radioactivity (when the rocks that the minerals are in became a solid) to the current level of radioactivity, and we have the age of the rock.

Although the moment at which a nucleus decays is unpredictable, a collection of atoms of a radioactive isotope decays exponentially at a rate described by a parameter known as the half-life, usually given in units of years when discussing dating techniques. After one half-life has elapsed, half of the atoms of the isotope in question will have decayed into a "daughter" isotope or decay product. In many cases, the daughter isotope itself is radioactive, resulting in a decay chain, eventually ending with the formation of a stable (nonradioactive) daughter isotope.

For most radioactive isotopes, the half-life depends solely on nuclear properties and is essentially a constant. It is not affected by external factors such as temperature, pressure, chemical environment, or the presence of a magnetic or electric field. So, the proportion of the original isotope to its decay products changes in a predictable way as the original isotope decays over time. This predictability allows the relative abundance of related isotopes to be used as a "clock" to measure the time to the present from the incorporation of the original isotopes into a material.

To make sure this clock is accurate, correlation between different isotopic dating methods may be required to confirm the age of a sample. For example, the age of very old rocks from western Greenland was determined to be 3.6 million years using uranium-lead dating and 3.56 million years ago using lead-lead dating, results that are consistent with each other.[11]

The precision of a dating method depends also on the half-life of the radioactive isotope involved. For instance, carbon-14 has a half-life of 5,730 years. After an organism has been dead for 60,000 years, so little carbon-14 is left that accurate dating is no longer possible. On the other hand, the concentration of carbon-14 falls off so steeply that the age of relatively young remains can be determined precisely to within a few decades. For much longer date spans, other isotopes must be used.

To summarize, radioactive dating is now the principal source of information about the absolute age of rocks and other geological features, including the age of fossilized life-forms and the age of the earth itself. So, thanks to the Book of Nature, we have come a long way in the right direction since the time James Ussher made his calculations.

[11] G. Brent Dalrymple, *The Age of the Earth.* (Stanford, CA: Stanford University Press, 1994), 142–143.

Part 2

Where Science and Religion Meet

The question of how God made the earth our home is ripe for much debate. There is a general conviction nowadays that science and religion are in conflict with each other. It is even believed that science has made religion irrelevant. If so, then the Big Bang, the development of galaxies, the formation of the earth, and the evolution of life has nothing to do with God. If so, then science and religion could nowhere meet. Is that true? The answer may surprise you.

Creation or the Big Bang?

Are we forced to make a choice between creation and the Big Bang? Or is there a way that both could be true? Let's see what each one entails.

The Big Bang

Modern science tells us that our universe most likely started with the Big Bang, some fourteen billion years ago. The English astronomer and mathematician Fred Hoyle is credited with coining the term *Big Bang* during a 1949 radio broadcast. Currently, the Big Bang theory is the prevailing cosmological model that explains the early development of the universe.

According to this theory, the universe was once in an extremely hot and dense state and then expanded rapidly. This rapid expansion caused the universe to cool and resulted in its present continuously expanding state. Don't take this expansion the wrong way — it's not that galaxies are moving out through space, but that the space between the galaxies is stretching.

How do we know all of this, given the fact that no one was around at the time? Edwin Hubble discovered in 1929 that the distances to far-away galaxies were generally proportional to their redshifts. Redshift occurs when an astronomical object is observed to be moving away from the observer, causing the emission or absorption features in the object's color spectrum to shift toward longer (red) wavelengths. Hubble's observation was taken to indicate that all very distant galaxies and clusters have an apparent velocity directly *away* from our vantage point; and the farther away, the higher their apparent velocity.

This phenomenon had already been suggested in 1927 by the Belgian priest, astronomer, and physicist Georges Lemaître of the Catholic University of Louvain. In 1931, Fr. Lemaître went further and suggested that the evident expansion of the universe, if projected back in time, meant that the further in the past, the smaller the universe was, until at some finite time, all the mass of the universe was concentrated in a single point—a "primeval atom," in Fr. Lemaître's words—where and when the fabric of time and space must have started.

Because the universe is expanding, looking out into space is like looking back in time. That is because light from objects that are far away takes longer to reach us than light from objects nearby. Cosmic distances are expressed in light-years. A light-year is the distance that light can travel in one year. Light moves at a velocity of about 300,000 kilometers per second, so in one year, it travels $9,500,000,000,000$ (9.5^{12}) kilometers. If an object is a million light-years away, the light from it that reaches us left that object a million years ago. Consequently, we see it as it looked some million years ago. Modern telescopes are so powerful that they can view objects many billions of light-years away, close to the time of the Big Bang. It has been found that the farther away galaxies seem to be (and the dimmer they appear to us), the greater is their redshift, and thus the faster they are moving away. This was the first direct evidence that the universe is not static (as most scientists had believed) but is expanding.

Thanks to the Big Bang theory, we can now raise the question of the beginning of time. Although it sounds improbable, modern science holds that time is something that began at one point. Albert Einstein had already showed us that both time and space are part of the physical world, just as much as matter and energy are. Since space-time is something that has not always existed, we might think that there must have been a "time" when there was no time and space—which cannot literally be true, of course. If one applies the laws of Einstein's "classical" general relativity, then there had to be a "point zero" at which space-time didn't yet exist. That point represents an outer limit to what we can know about the universe, since it makes no sense, and is self-contradictory, to speak of what happened before the beginning of time, taken in a temporal sense.

What Is Creation?

At the beginning of this book, we found out that creation is not a matter of six days or even of six extended periods of time. Creation in itself is not an *event* at all. It is not something that happened at a certain place (such as Paradise) or at a certain time (not even long ago). In a nutshell, creation happened not "*at* the beginning" of the world but "*in* the beginning." The first line in the book of Genesis says it clearly, "*In the beginning*, God created the heavens and the earth." "In the beginning," however, is not something that happened *before* anything else—something like "first thing in the morning."

Why can creation not have happened at a certain time? The reason is simple. Creation cannot follow a timeline, as time itself has to be created to begin with. Time cannot exist before there was time, because there can be no time before there was time. Time as well as space are part of the physical world. They are part of something that must exist already, the physical world; otherwise there wouldn't be time either. Even St. Augustine knew this long ago, when he said, "There can be no time without creation."[12] He did not need science to come to this conclusion.

In other words, the first words of Genesis—"In the beginning"—do not refer to a trigger event at the beginning of time (something like a Big Bang), because there could not be any events (neither a Big Bang nor evolution) without creation. We cannot even say that creation must come first—that is, before anything else can follow—for there is no "first" or "before" until time has been created. How, then, should "In the beginning" be understood? Not as the beginning of everything in a temporal, chronological, or sequential sense, but rather in a transcendent sense—that is, in the sense of "originally" (the basis of any beginning); that is, the origin of everything in the universe, in which everything else finds its foundation.

When we say that God created the world, we are saying that everything that exists depends on God; He created everything; He is the origin of all there is; He is the source of all there is; He is the foundation of all there is. Take the analogy of a novel: the *beginning* of the novel consists of its first words or sentences; but the *origin* of the novel is found in its author. No matter what the first lines of the novel are, there must be an author. Well, creation is about the *origin* of the universe; science is about the *beginning* of the universe. Obviously, the unfolding of the universe, starting with the Big Bang, is a process that plays out in time and can be studied by the physical sciences, which is something we did in the previous chapters. Creation, on the other hand, creates chronology, but it is not a part of chronology.

So, what is creation then, if it's not an event at a certain place and at a certain time? The answer is a bit philosophical, so please

[12] Augustine, *Confessions*, 11 13.

bear with me. The basic idea is that nothing can create itself. Nothing can just pop itself into existence; it must have a cause, because it does not and cannot have the power to make itself exist. As the saying goes, "Nothing comes from nothing." For something to create itself spontaneously, or produce itself spontaneously, it would have to exist before it came into existence—which is logically and philosophically impossible. Since nothing can create itself, we need creation. Many others have confirmed this. The theologian David Bentley Hart, for instance, puts it this way: "Physical reality cannot account for its own existence for the simple reason that nature—the physical—is that which by definition already exists."[13] Or, in the words of particle physicist Stephen Barr, "Anything whose existence is contingent (i.e., which could exist or not exist) cannot be the explanation of its own existence."[14]

So, the explanation for all that exists has to be found somewhere else—in creation, that is! If rightly understood, creation is not an event at all, not a "one-time deal," but it answers questions about where this universe ultimately comes from, how it came into being, and how it stays in existence. The answer to these questions is that the universe, and everything in it, does not come from the Big Bang, but may have started with the Big Bang. The universe is ultimately something that finds its origin and foundation somewhere else. Without creation, there could not be anything—no Big Bang, no gravity, no evolution, not even a timeline. Creation sets the stage for all these things and keeps all of them in existence.

To explain this a bit further, I would like to introduce an important distinction that St. Thomas Aquinas, a Doctor of the Church,

[13] David Bentley Hart, *The Experience of God: Being, Consciousness, Bliss* (New Haven, CT: Yale University Press, 2014), 96.

[14] Stephen M. Barr, "Modern Physics, the Beginning, and Creation: An Interview with Physicist Dr. Stephen Barr," *Ignatius Insight*, September 26, 2006.

uses. He distinguishes between producing (*facere*) something and creating (*creare*) something.[15] Most people use these two terms interchangeably, but Aquinas advises us clearly to separate them. Science concerns itself with the production of something from something else; it is about changes taking place in this universe. Creation, on the other hand, is about the coming into being of something out of nothing—which is not a change at all, and certainly not a change from "nothing" to "something."

In other words, the Creator doesn't just take existing stuff and fashion it, as does the Demiurge in Plato's *Timaeus*. Nor does the Creator use something called "nothing" and then create the universe out of that. Rather, God calls the universe into existence without using existing space, matter, time, or anything else. In 1215, the Fourth Lateran Council taught that the universe was "created out of nothing" [*creatio ex nihilo*] at the beginning of time.[16] More on this later.

We need to keep stressing that *creating* something "out of nothing" is not *producing* something out of nothing—which would be a dangerous misconception, for it treats "nothing" as some kind of something. In contrast, the Christian doctrine of creation "out of nothing" (*ex nihilo*) claims that God made the universe out of no thing, without making it out of any thing. Creation has everything to do with the philosophical and theological question of why things exist at all, before they can even undergo change. Speaking to God, the book of Wisdom asks, "How could a thing remain, unless you willed it; or be preserved, had it not been called forth by you?" (11:25). To use a phrase of Shakespeare, creation is about "to be or not to be."

Creation is not something that happened long ago, nor is the Creator someone who did something in the distant past, for the Creator does something at all times—by keeping our contingent

---

[15] *De Symbolo Apostolorum*, 4, 33.
[16] Denzinger, "*Enchiridion*," 428 (355)

world in existence. Whereas the universe may have a beginning and a timeline, creation itself does not have a beginning or a timeline; creation makes the beginning of the universe and its timeline possible. William E. Carroll is right when he stresses that we should never confuse temporal *beginnings* with theological *origins*. In his words, "We do not get closer to creation by getting closer to the Big Bang."[17] In other words, we do not get closer to our origin by getting closer to the beginning of this universe. Once we lose sight of this important distinction, we are in for a serious mix-up with dangerous consequences.

Creation and the Big Bang

Many scientists think that the discovery of the Big Bang has made the religious idea of creation and "creation out of nothing" obsolete. The late astrophysicist Stephen Hawking, for instance, talked about the Big Bang in terms of what he calls a "spontaneous creation" when he said, "Because there is a law such as gravity, the Universe can and will create itself from nothing. Spontaneous creation is the reason there is something rather than nothing."[18] This, in turn, made the late astrophysicist Carl Sagan exclaim, in the introduction to Hawking's *Brief History of Time*, that such a cosmological model has "left nothing for a creator to do." The cosmologist Lee Smolin also made sure there is no space left for a Creator by proclaiming that "by definition the Universe is all there is, and there can be nothing outside it."[19] And cosmologist Alexander Vilenkin has a

[17] William E. Carroll, "The Genesis Machine: Physics and Creation," *Modern Age* (Winter–Spring 2011).

[18] "Stephen Hawking and Leonard Mlodinow, *The Grand Design*," *Times Eureka* 12 (September 2010), 25.

[19] Lee Smolin, *Three Roads to Quantum Gravity* (New York: Basic Books, 2001), 17.

universe suddenly pop into existence from "nothing" by a so-called quantum fluctuation—a fluctuation of a primal vacuum.[20] Or take the physical chemist Peter Atkins, who claims that science has the power to account for the "emergence of everything from absolutely nothing."[21] What they all seem to agree on is that we have here the physical explanation for what religious people used to call "creation out of nothing."

These men are certainly well schooled in science, but typically they are not knowledgeable in philosophy and theology, as we will soon find out. They tend to think that terms used in science have the same meaning as terms used in philosophy, metaphysics, and theology. Because of this confusion, they also tend to think that science can answer philosophical and theological questions—questions such as "What is creation out of nothing?" That is a very dubious assumption, as we will discuss next.

Hawking's claim that the universe can create itself out of nothing is very hard to defend. Indeed, his idea of a "spontaneous creation" is sheer philosophical magic. How could the universe create itself out of nothing—not to mention cause itself? The law of gravity cannot do the trick, because before the universe could ever create itself (if even possible), there would have to be laws of physics—which are ultimately the laws that govern the existing, created universe.

So, Hawking and his followers are saying that the laws that govern an existing universe can generate that universe and bring it into existence, along with its laws of nature, all by themselves before either exists—which makes for a logical contradiction. Hawking's explanation is a perfect example of circular reasoning: How can

[20] Alexander Vilenkin, *Many Worlds in One* (New York: Hill and Wang, 2006), 185.
[21] P. W. Atkins, "The Limitless Power of Science," in *Nature's Imagination*, ed. John Cornwell (Oxford: Oxford University Press, 1995), 123.

gravity exist if there is no universe? And if there is no universe, how can gravity be the reason for the creation of the universe? To have anything—a universe, a law of gravity, anything—you have to have creation to begin with.

To claim, as Smolin does, that the universe is all there is, with nothing outside it, is very hard to defend, too. How could we possibly know that the universe is all there is, unless we define ahead of time that the universe is all there is ? This is another example of circular reasoning, for it begins with what it is trying to end with. Besides, it is very hard, if not logically impossible, to conclude that there is nothing outside the universe. It is impossible to close a search for something with the conclusion that what we searched for is *not* there. No searches ever conclusively reveal the *absence* of their object—just keep searching! Absence of evidence is not evidence of absence.

To claim, as Atkins does, that science has the power to account for the "emergence of everything from absolutely nothing" is very hard to defend as well. St. Thomas Aquinas would keep hammering on the distinction between producing and creating, or between changing and creating, or between something and nothing. Creation does not mean changing a "no thing" into a something or changing something into something else. Creation means bringing everything into being and existence. To be or not to be: that's indeed the question for philosophy and theology, but not for science.

Whenever something changes in the material world, there must be something that changes, for nothing comes from nothing. When God creates something out of nothing, this "nothingness" in philosophy is not a highly unusual kind of exotic "stuff" that is more difficult to observe or measure than other things; it is not some kind of element that has not found a position yet in the Periodic System; it is in no way a material thing at all that can change into something else, but is actually the absence of anything.

In the Beginning

To claim, as Vilenkin does, that the universe popped out of "quantum tunneling from nothing," as a fluctuation of a primal vacuum, is also very hard to defend. The problem with his claim is that the term *vacuum* in physics is not the same as the term *nothing* in philosophy and theology. When an electron and a positron collide, they can annihilate and thus change into "nothing" (*nihil*). What really happens when they annihilate is that they emit a burst of energetic photons—which is certainly not nothing in a metaphysical sense. The reverse can occur, too; this happens when an "empty" space is filled with an electric field, but no particles are in it. In that situation, there is a certain probability that suddenly an electron-positron pair will pop out of "empty" space. It happens by a process called "quantum tunneling," which causes some system to change from one state (an electric field without particles) into another state (a changed electric field with two particles). These are different states, but of the same system. It must be realized, however, that a pair of particles does not suddenly appear out of nothing but emerges from an electric field in an existing system—which, again, is not nothing in a metaphysical sense.

In this context, Stephen Barr uses the analogy of having a bank account with no money in it: even if we have "nothing" in the bank, we still have a bank account, with all that comes with it, but it happens to be in a "no-money state" for us. This kind of "nothing" is different from having no bank account at all.[22] Characterizing a vacuum fluctuation as being the scientific equivalent of *creatio ex nihilo* entirely misconstrues the discussion. As John Polkinghorne puts it, "A quantum vacuum is not *nihil*, for it is structured by the laws of quantum mechanics and the

---

[22] Stephen M. Barr, *Modern Physics and Ancient Faith* (Notre Dame, IN: University of Notre Dame Press, 2003), 277–278.

equations of the quantum fields involved." William E. Carroll characterizes a primal vacuum succinctly when he says, "It is still something—how else could 'it' fluctuate?"[23] We are dealing here with a system that has a set of possible states. Obviously, a state is a state, not nothing. It is a specific state of a specific, complicated quantum system governed by definite laws. In physics, *nothing* is really something, whereas in philosophy and religion, it refers literally to "no thing."

For all the above reasons, it is very hard to defend the claim, as so many scientists do, that the universe just popped into existence out of nothing. The kind of "nothing" scientists talk about is not the kind of "nothing" philosophers and theologians speak of when they use the expression "creation out of nothing." In other words, the term *vacuum* of physicists is not the same as the term *nothing* in philosophy and theology. The term *nihil* in physical annihilation is not the same as the term *nihil* in *creation ex nihilo.* Therefore, to claim, as too many do, that the religious idea of "creation out of nothing" is obsolete is very hard to defend.

It is a mistake—a category mistake, if you will—to use arguments from the natural sciences to deny creation in philosophy, metaphysics, and theology. Some scientists have even used the second law of thermodynamics—which states that energy can never be created or destroyed—to prove that energy could never have been created in a theological sense. That law, however, applies only to an existing universe and cannot apply to a universe that has not come into existence yet. The problem is again that *creating* in science is *producing,* not creating out of nothing by the Creator.

[23] William Carroll, "Thomas Aquinas and Big Bang Cosmology," Jacques Maritain Center, https://maritain.nd.edu/jmc/ti/carroll. htm.

In the Beginning

So, from now on, we should answer *why* questions with two kinds of answers at two levels—a lower level and a higher level. Why are there galaxies? The answer at the lower level is: because of the Big Bang. The answer at the higher level is: because of "creation out of nothing." Why are there planets? The answer at the lower level is: because of the galaxies. The answer at the higher level is: because of "creation out of nothing." Why are there living beings? The answer at the lower level is: because of the evolution life went through on Planet Earth. The answer at the higher level is: because of "creation out of nothing." Different types of questions call for different types of answers.

The fact remains that science is not about nothing; it is always about something, such as galaxies, planets, and organisms. Science cannot fill the immense chasm called "nothing." Only "creation out of nothing" can. That's why the question "Why is there something rather than nothing?"[24] remains pressing. Why is there gravity? Why are there laws of nature? Why is there a Big Bang? Why are there galaxies? Why is there a planet called Earth? Why is there continental drift? Why is there life on Earth? Why are we here? These are questions that can be answered at level one but should also be answered at level two. More in general, we should ask at level two, "Why is there something rather than nothing?" Why not just nothing rather than something? The only reasonable answer is: because there was *creation out of nothing.*" The idea that there is nothing to explain why there is something rather than nothing is absurd—it takes us back to nothing. So, we cannot avoid that momentous question "Why is there something rather than nothing?" The answer cannot be that the universe can create itself out of nothing. There has to be something or someone outside or beyond the universe.

[24] G. W. Leibniz, *The Principles of Nature and Grace, Based on Reason* (1714).

Who Is the Creator?

Once we realize that the Big Bang still requires creation, the next question is "Who is the Creator?" Of course, the Creator is God. But this just shifts the question from "Who is the Creator?" to "Who is God?" Let's find out first what God is *not*.

First, God is *not* part of a pantheon—a collection of gods and goddesses, deities and divinities. In a pantheon, even if one of the deities may have a special status—first among equals, so to speak—ultimately, they are all of the same caliber. At best there is a "supreme being" among them, some deity such as Zeus in Greek mythology. We see something similar with Brahman in Hinduism.

It is a world of polytheism that the Apostle Paul stepped into when, around 51, he went to Athens in Greece, where he found idols everywhere. But he also saw an opening for Christianity there: "For as I walked around looking carefully at your shrines, I even discovered an altar inscribed, 'To an Unknown God.' What therefore you unknowingly worship, I proclaim to you" (Acts 17:23). St. Paul started a dialogue with polytheism in order to promote Judeo-Christian monotheism.

Second, God is *not* a superbeing with superpowers. All the gods in the pantheon are eerily similar to human beings, yet what makes them "gods" is that they have powers that exceed human powers: they have "superpowers"—whatever that means. They can control weather, thunderstorms, earthquakes, the harvest—you name it. They have powers raised to the power of a zillion. It is these powers that make people adore and venerate them. Their powers, however, are as limited as ours, even if raised to the zillionth power.

Third, God is *not* a hypothesis of science. Mesmerized by the success stories of science and the scientific method, many people praise science for its ability to submit hypotheses to tests, either in the laboratory or in the field. God cannot be a hypothesis, however. First of all, God is not a spatiotemporal entity like all the objects of science. God is an infinite power outside and beyond the spatiotemporal dimensions of our universe. God did not create the world in time, but with time — or having time in it. He did not create the world in space, but He created space in the world — or a world full of space. From out of His very Being, God gives being to the whole world and all things in it, including time and space. That's the first point where science can come in — but no sooner.

Besides, it is very questionable to demand indisputable evidence of God's existence as we do in science. Catholic philosophers such as Peter Kreeft point out that this demand for scientific evidence through something like laboratory testing is in effect asking God, the Supreme Being, to become our servant. Kreeft and other philosophers argue that the question of God should be treated differently from other knowable objects in that this question regards not that which is *below* us, but that which is *above* us. Such a stand raises the discussion to a higher level.

Fourth, God is *not* a "god of the gaps." Even those who believe strongly in the power of science must acknowledge that we are still not able to explain everything in life — there are large gaps in our scientific knowledge. Therefore, it remains a timeless temptation for some to invoke God as an explanation for what we cannot scientifically explain at this moment, or perhaps ever. The god some use to explain the gaps in our knowledge has become known as the "god of the gaps" — a god who uses divine interventions to make up for gaps in his creation. Making God a "repairman," however, puts Him on the same level as all the other factors and causes operating in this universe. In that way, God has been downgraded to someone like Zeus working *inside* the universe.

The problem with invoking divine interventions is that science keeps making progress, and these new discoveries may make direct divine interventions unnecessary. If, in fact, the frontiers of knowledge are being pushed further and further back — which is bound to be the case—then God is being pushed back with them and is therefore continually in retreat. God seen as a "god of the gaps" can serve only as a provisional explanation for what science has not yet explained. Cardinal Avery Dulles once wisely remarked, "As a matter of policy, it is imprudent to build one's case for faith on what science has not yet explained, because tomorrow it may be able to explain what it cannot explain today."[25]

Fifth, God is *not* a delusion of wishful thinking. Sigmund Freud, for instance, believed that belief in God is a reversion to childish patterns of thought in response to feelings of helplessness and guilt. We supposedly feel a need for security and forgiveness, and so we invent a source of security and forgiveness: God. Religion is thus seen as a childish delusion, whereas atheism is taken as a form of grown-up realism.

It is very doubtful, however, that someone such as Freud really did refute God's existence as an illusion or a delusion. First of all, if Freud claims that basic beliefs are the rationalization of our deepest wishes, wouldn't this entail that his own atheistic beliefs could also be the rationalization of his own wishes? Besides, even if belief in God were wishful thinking, one could never prove that it is nothing more than wishful thinking. The God whom one would like to exist may actually exist, even if the fact that one wishes it may encourage suspicion. In addition, the claim that the brain created God could easily be countered with the reversed claim that it was God who created the brain. If that's so, we do not project a human image onto Heaven, but Heaven shows us what we as human beings can and should be like.

If God is *none* of the above, who, then, is God, the Creator? What we do know about God and His creation comes first and

[25] Avery Cardinal Dulles, "God and Evolution," *First Things* (October 2007).

foremost through this universe. But how do we then get from what the universe tells us about creation to the Creator of this universe? A person who can help us enormously to make this giant step is St. Thomas Aquinas. (Sorry, it's going to be a short high-brow talk, but I need it for the rest of this book.)

Aquinas knew very well how things cause each other. He said that whenever something undergoes change, something else must be causing that change; nothing can be the cause of its own change. Therefore, whenever something changes, this change must have been brought about by something other than itself. As nothing can cause itself to exist, so nothing can cause itself to change. The structure of the universe is such that all causes have effects—with "like causes having like effects."

So, all things in this universe have a cause-and-effect relationship. Aquinas uses the example of fathers begetting sons.[26] A father begets a son, who begets a son, who begets a son, and so forth. Without the first father begetting his son, the last son would not exist and therefore could not continue the series of fathers and sons. Yet, if the first father dies, that doesn't prevent the last son from begetting yet another son. This is an *accidental* series of causes, in which the earlier causes need no longer exist in order for the series to continue. Nowadays, we could translate this to science and say that the Big Bang caused galaxies; galaxies caused Planet Earth; Planet Earth caused life; and so forth. But the Big Bang is no longer necessary for the sequence to continue.

Aquinas, however—and this is the salient point—also distinguishes an *essential* series of causes, in which the first, and every intermediate member of the series, must continue to exist in order for the causal series to continue as such. To explain how different an essential series is from an accidental series, I would like to

---

[26] Thomas Aquinas, *Summa Theologica*, I, q. 46, art. 2, ad. 7.

introduce an example that the philosopher Edward Feser uses: a cup of coffee sitting on someone's desk.[27] The cup has no capacity on its own to be three feet from the ground; it will be there only if something else, such as the desk, holds it up. But the desk, in turn, has no power of its own to hold the cup there. The desk, too, would fall to the earth unless the floor held it aloft. And the floor, for that matter, can hold up the desk only because it is itself being held up by the house's foundation, and the foundation by the earth, and, in turn, the earth by the structure of the universe. Something similar could be said about a book sitting on the shelf of a bookcase. All these "intermediaries" keep each other in tow. None of these things, however, could hold up anything at all without something that holds them up having to be held up itself.

That ultimate "something" is what Aquinas calls the *Primary Cause*, known more colloquially as the *First Cause*. Without it, nothing in an essential series of accidental causes could really be explained. This First Cause has the power to produce its effects without being caused by something else. It has *inherent* causal power, whereas all other causes have only *derived* causal power. In other words, the need for causes must come to an end: there must be or have been a cause that is not itself in need of a cause — a First Cause, that is. Ultimately, it is God who holds a cup or a book in its place. Without God, they would collapse, or would not even be there. Without God, nothing in the universe would have an ultimate, firm foundation.

Causes that depend for their existence on the Primary Cause are called *secondary* causes by Aquinas. Thanks to the First Cause, the secondary causes can be causes of their own. Without the First Cause, there could not be any secondary causes. Again, creation is

---

[27] Edward Feser, *Five Proofs of the Existence of God* (San Francisco: Ignatius Press, 2017), 22ff.

very different from what science is searching for. Whereas science deals with secondary causes, creation is about the First Cause—that is, the Creator.

As Pope Benedict XVI put it, "Thomas observed that creation is neither a movement nor a mutation. It is instead the foundational and continuing relationship that links the creature to the Creator, for he is the cause of every being and all becoming."[28] The *Catechism of the Catholic Church* puts it as follows, "God is the first cause who operates in and through secondary causes.... 'Without a Creator the creature vanishes.'"[29]

Thanks to the distinction between the First Cause and secondary causes, we can better distinguish between God, the Creator, and what the Creator created. God is the First Cause, but the laws of nature and their effects are secondary causes. God's divine causality makes the natural causality of the universe possible. The latter depends on the former, but they both operate in their own ways. More importantly, they operate on two completely different levels. The physical causality of nature, as studied in science, reigns "inside" the universe, linking causes in a chain or network of secondary causes. God, on the other hand, reigns from "outside" the universe as a Primary or First Cause, thus providing somehow a point of suspension, so to speak, for the network of secondary causes. What makes the Primary Cause so unique is that it needs *no* cause. And what is so special about secondary causes is that they *do* need a cause.

In other words, God, as a First Cause, is not a supercause among other causes; He is above and beyond all secondary causes, bringing

[28] Benedict XVI, Address to the Members of the Pontifical Academy of Sciences, October 31, 2008.
[29] *Catechism of the Catholic Church* (CCC), no. 308, quoting *Gaudium et spes*, 36 § 3.

them into being and keeping them in existence. God lets them do their own work—secondary causes, that is. Without a Primary Cause, there could not be a chain of secondary causes. It is like the framework around a spiderweb; without that framework, the web could not exist. In a similar way, God is the Eternal, Infinite Cause, "in whom we live and move and have our being" (Acts 17:28).

It is classically said that the Primary Cause is uncaused, not self-caused, the source of all being; not a cause prior to and larger than other causes, but a Primary Cause; not a power stronger than and superior to all other powers, but an Infinite Power; not some superbeing among other beings, who acts like other beings, but an Absolute Being.

When Aquinas described God as the First Cause, he meant "first" not merely in the sense of being before the second cause in time, and not in the sense of coming before the second cause in a sequence, but rather "first" in the sense of being the *source* of all secondary causes, having absolutely primal and underived causal power—a power from which all other causes derive their causal powers.

Seen in this light, a force such as gravity is a secondary cause. By allowing an "inferior" cause like this to operate, God does not have to be the direct cause of every stone falling to the ground—for that would make Him a secondary cause. We do not have to wonder about God's will every time a stone falls, even if it strikes us on the head. God has given us a secondary cause—the force of gravity—which is the direct cause of each stone's earthly plummet.

Let me use a simple example to make this distinction more tangible. When I cut wood with an ax, I cut the wood, but so does the ax. When cutting wood, I am somehow a "primary cause," and the ax acts like a secondary cause. I, as a primary cause, use the ax as a secondary cause—and together we make things happen.

When we talk about God in the Judeo-Christian tradition, we are talking about this God, whom Aquinas called the First Cause.

The *Catechism* confirms this, "Man can come to know that there exists a reality which is the first cause and final end of all things, a reality 'that everyone calls God.'"[30]

You might wonder, though, whether this is really all we can say about the Creator. Fortunately, the Bible tells us much more about this God than human reasoning can achieve by itself. The God who is the Primary Cause turns out to be also a God of Love, who provides for His creatures and who came down from Heaven through the Incarnation of His Son. So, God is much more than what philosophy can tell us. But at least God is not less than what philosophy can reveal to us when we use the power of reason that comes from God, too.

Let's leave it at that for now.

---

[30] CCC 34, quoting Thomas Aquinas, *Summa Theologica*, I, q. 2, art. 3.

4

What Comes with Creation?

Without God, secondary causes could not exist. Creation brings secondary causes into existence and keeps them in existence. There is more that comes with creation, however: physical laws and physical constants. Let's find out what that means.

Physical Laws of Nature

Much of what happens in our world is determined by the laws of physics, chemistry, and so on — more collectively called the *laws of nature*. Let there be no misunderstanding: laws of nature have always existed; they were discovered, not invented. The simplest, and arguably the oldest, law of physics is Archimedes's principle (ca. 250 B.C.), which he formulated as follows: "Any object, wholly or partially immersed in a fluid, is buoyed up by a force equal to the weight of the fluid displaced by the object."[31]

The best-known law of nature in physics is probably the law of gravity, first discovered by Isaac Newton (ca. 1686). This law played a pivotal role in the development of the universe and the

---

[31] *The Works of Archimedes*, ed. Sir Thomas Heath (Dover Publications, 2002), 257.

formation of the earth, as we will see. The law is rather deterministic and limits the effects of random events dramatically. The force of gravity obeys an "inverse square law." The pull of gravity between two objects is proportional to the masses of the objects and inversely proportional to the square of the distance between their centers of mass. So, if the distance were doubled (x 2), then the gravitational pull would be only one-fourth as great ($1/2^2$ = 0.25). With these laws of motion we can accurately compute the trajectories of planets around the sun and the motion of stars.

There are many more laws—for example, Newton's three laws of motion.

- First law: a body at rest will remain at rest, and a body in motion will remain in motion, unless it is acted upon by an external force.
- Second law: the net force acting on an object is equal to the mass of that object times its acceleration.
- Third law: for every action, there is an equal and opposite reaction.

These laws of motion are needed to understand the motion of objects in our universe. Using just a few equations, for instance, scientists can describe the motion of a planet and forecast eclipses of the moon.

Newton did some important groundwork, but later developments in physics added many more laws of nature. With this deepening of physics came a more unified picture of the universe, relating phenomena as seemingly diverse as heat, light, magnetism, and gravity.

There are still many challenges left. One of them is the search for what unites the three nongravitational forces of nature—the electromagnetic force, the weak nuclear force, and the strong nuclear force (more on this later). Physicists are still waiting for what they call a "grand unified theory" to answer such questions.

What Comes with Creation?

Unlike the many laws of nature that are deterministic, some laws in physics are inherently probabilistic—especially since the development of quantum physics. The best-known probabilistic law in physics is the law of radioactive decay. Although we do not know whether an individual isotope atom will decay and when, we do know what happens in the aggregate, based on the statistical law of large numbers. Take, for instance, radium, with its 88 protons. Its most stable isotope, radium-226, may decay, without any outside interference, by emitting a so-called alpha particle, which has 2 protons and 2 neutrons. This changes radium-226 (with 88 protons) into radon-222 (with 86 protons). Even though the half-life of this radioactive isotope can be determined (1,600 years), nothing determines when one particular atom will disintegrate—even though we can register the individual decay event on a Geiger counter.

We are dealing here with laws of nature—whether deterministic or probabilistic. These laws are part of creation. Whereas the Creator creates something out of nothing, the universe produces something out of something else—by having planetary motions follow physical laws and by having evolutionary developments follow biological laws. God does not make things Himself—in a manual and interventional way, so to speak—but He makes sure they are being made through His laws.

As we said earlier, the laws of nature that govern an existing universe can never generate that universe and bring it into existence—that amounts to a logical contradiction. Neither can the universe generate its own laws of nature, for that would require other laws of nature. The laws of nature cannot bring themselves into existence and cannot explain themselves. In other words, the laws of nature come from somewhere else: creation.

No wonder that some scientists dislike, or at least avoid, the term laws of nature. It reminds them too much of laws implemented by a higher authority, a lawgiver. Instead, they often describe laws of nature as mere descriptions of certain regularities in the universe. Even if we declare the laws of nature pure descriptions of cause-and-effect relationships, however, we are still left with the question as to where this regularity comes from. So, ultimately, we cannot evade the idea of a Lawgiver.

Physical Constants

In addition to the *laws of nature*, much of what happens in our universe is determined by so-called *fundamental physical constants*. These are physical constants whose value cannot be derived from something else—at least not at this time; they can be determined only by physical measurement. This is one of the unsolved problems of physics, because the numerical values of these constants are not understood in terms of any widely accepted theory. Perhaps someday we will be able to derive their value from a more general theory.

There are four basic forces in nature: gravity, electromagnetism, weak nuclear interactions, and strong nuclear interactions. Let's focus first on the strong nuclear force, which holds most ordinary matter together because it binds neutrons and protons to create atomic nuclei.[32]

Then there is the weak nuclear interaction force, which is the interaction between subatomic particles (such as quarks and electrons). This force causes radioactive decay and thus plays an essential role in nuclear fission. The weak interaction takes place only at very small, subatomic distances, less than the diameter of a proton. The force is, in fact, termed "weak" because its field

---

[32] Its value is 2.22457 milli-electron volts (MeV).

strength is typically several orders of magnitude less than that of the strong nuclear force or electromagnetic force.[33]

Another example of a physical constant is the mass of the proton.[34] Protons are very stable, whereas an isolated neutron is very unstable and quickly disintegrates into other types of particles. The key for the difference in stability of protons and neutrons is located in their mass. Mass (m) is related to energy (E): $E = mc^2$. A proton is a seventh of a percent lighter than a neutron, so it has a bit less energy packed, not enough to fall apart. If the proton's mass were even 1 percent larger than the neutron's, however, then protons would be unstable. This would make hydrogen-1 (with only one proton and no neutron) unstable, so there would be no ordinary hydrogen in the universe, and therefore no water and no organic molecules — and we wouldn't be here either.

And then there is the electromagnetic fine-structure constant,[35] which is another pure, given number that is at the basis of all the structural activity of biology, physics, and chemistry. We won't go into more detail.[36]

As said before, a long-sought goal of theoretical physics is to find first principles from which all the fundamental constants, such as the electromagnetic fine-structure constant, can be calculated and compared with the measured values, but so far, this goal has remained elusive. At this point, the values of these constants seem to be a given in the fabric of this universe.

What can we learn from all of this? Although randomness does play a role in the history of the universe, all the aforementioned

---

[33] The weak interaction has a value of 0.000001, compared with the strong interaction's coupling constant of 1, whereas the electromagnetic force has a coupling constant of 0.00729927.

[34] Its value is 938.272 MeV.

[35] It has a value of about 1/137.036 which is close to 0.007297353.

[36] See Barr, *Modern Physics and Ancient Faith*, 119–137.

physical constants and physical laws of nature place an enormous constraint on all that seems to take place so "randomly" in the development of the universe and the earth.

The question remains, of course, as to where the specific values of these constants came from. Some of them, or perhaps even all, may in time be derived from a higher-level law. For instance, at one time, the boiling point of water was taken as a physical constant, but it is now considered the result of quantum mechanical laws. In time, the speed-of-light constant may be tied to Planck's constant defining the smallest possible unit of space and time.

Even if all constants were ever to be derived from a single unified theory, however, some hoped-for supertheory, then the next question would be, of course, where that supertheory came from. We will get to that later.

Anthropic Coincidences

As we found out, fundamental physical constants often have very specific values, and physical laws of nature often have equations with terms in them that also have specific values. That could be seen as a "coincidental" outcome, in that the laws of physics seem to coincide exactly with what is required for the universe to be able to produce life, including intelligent beings like us, with reason and free will. What is even more striking is the fact that these values often have to be of a very specific magnitude. Had they been different, the universe would have had a very different outcome and would probably not have been able to launch life, let alone human life. For that reason, they are often called "anthropic coincidences" (from the Greek word *anthropos*, meaning "man").

Let's begin with the law of gravitation, which plays a role in most of what we are discussing in this book. Everything in the universe, from the tiniest atom in a gas to the largest planet, moves in a precise trajectory governed by the exact mathematical equations of the laws of nature.

If the strength of gravity were not based on the inverse square of the distance ($1/distance^2$—or $1/d^2$), the orbits of the planets in the solar system would be much more complicated than ellipses and would, in most cases, not form closed curves at all. William Paley, in 1802, pointed out that if the law of gravity had not been an inverse square law, the earth and the other planets would not be able to remain in stable orbits around the sun.[37] Instead, there is a balance between the force of gravity and centrifugal force. Otherwise planets would fly off to infinity or plunge into the stars they were orbiting.

Paley also remarked that if space had not been three-dimensional but, for example, four-dimensional, gravity would have decreased as the inverse cube of the distance ($1/d^3$), rather than the inverse square law ($1/d^2$). That change in the character of gravity would also have made planetary orbits unstable, and the continued existence of the solar system would have been in constant jeopardy. We take it for granted that space is three-dimensional, but it does not have to be that way; it's not a logical or metaphysical necessity, but an empirical fact. Had that fact been different, it would have been impossible for planets to orbit stably around stars: they would either have plummeted into stars or flown off into space.

On the other hand, had there been fewer than three space dimensions, as Stephen Barr notes, complex organisms would have been impossible for quite a different reason. Complex neural circuitry, as is needed in a brain, would not be possible in two dimensions. If one tries to draw a complicated circuit diagram on

---

[37] William Paley, *Natural Theology*, 12th ed. (London: J. Faulder, 1809), chap. 22, 390–391.

a two-dimensional surface, one finds that the wires must intersect each other many times, which would lead to short circuits.[38]

Let's discuss next what the consequences would be if certain physical constants—some of which we discussed earlier—had been smaller or larger than they are.[39] Many consider these constants to be the way they just happen to be, but, as I indicated earlier, because they turn out to be favorable for human life, they are often called anthropic *coincidences*.

Let's start with the strong nuclear force—the force of nature that cements atomic nuclei together. What would happen if the strong nuclear force were stronger or weaker than it is? If it were only 10 percent *weaker*, it would choke off the process of making the elements at the very first step, for it would be too weak to hold the nucleus of hydrogen-2 together. Obviously, all the next steps could not occur then. It would be impossible to make a living being out of just hydrogen. If, on the other hand, the strong nuclear force had only been a few percent *stronger* than it is, an opposite disaster would have occurred. It would have been too easy for hydrogen nuclei to fuse together, so the nuclear burning in stars would have gone way too fast. The entire process would have been too short for more complex forms of life such as ours to emerge.

Another example is the strength of the electromagnetic force, which is controlled by the fine-structure constant.[40] This force is some one hundred times weaker than the strong nuclear force. Whereas the strong nuclear force tries to hold the nucleus together, the electromagnetic force tries to blow the nucleus apart, for the protons exert an electrical repulsion on each other. If the nucleus

[38] Barr, *Modern Physics and Ancient Faith*, 133.

[39] Paul Davies, *The Accidental Universe* (Cambridge, UK: Cambridge University Press, 1982).

[40] It is close to 0.0072992701.

has enough protons in it, the electrical repulsion of the protons will overpower the nuclear force that holds the nucleus together and will blow the nucleus apart. This means that a nucleus can contain only a limited number of protons to remain stable. Since the electromagnetic force is about one hundred times weaker than the nuclear force, we reach this limit at around one hundred protons in the nucleus. As a matter of fact, the largest nucleus naturally occurring in the universe—that of uranium—has ninety-two protons.

Had the electromagnetic force been much *smaller*—say, only one-fifth of what it is—then there would have been only some twenty-five elements in nature. Some elements essential for human life, such as calcium and iron, would not have been available. On the other hand, had the electromagnetic force been *larger*, the electrical energy packed inside a hydrogen nucleus would have been so great as to make it unstable. The weak interaction would then have made all the hydrogen in the world decay radioactively, with a very short half-life, into other particles. The world would have been left devoid of hydrogen, and therefore of water, necessary for life.

Another constant, required for cosmology, is the cosmological constant of Einstein's equations for general relativity.[41] When positive, this constant acts as a repulsive force, causing space to expand; when negative, it acts as an attractive force, causing space to contract. If it were too *large*, space would have expanded so rapidly that galaxies and stars could not have formed, and if too *small*, the universe would have collapsed before life had had a chance to evolve.

These examples of constants are probably the best examples of anthropic coincidences. There are many other constants in physics, but they may not affect the way our universe developed. One of them is the speed of light.[42] This value (denoted by c) is used, for

---

[41] Its value is approximately 10^{-120}.

[42] Its exact value is 299,792,458 meters per second.

instance, in the famous equation $E = mc^2$. Important as the value of this constant is, it would not be called an anthropic coincidence.

What all the anthropic coincidences have in common is that their numerical values are not understood in terms of any widely accepted theory. As Stephen Hawking has noted, "The laws of science, as we know them at present, contain many fundamental numbers, like the size of the electric charge of the electron and the ratio of the masses of the proton and the electron."[43] Some scientists see the values of these physical constants as "coincidences" — they just happen to be this way and just happen to be favorable to the development of human life. But that may need to be seen.

The Anthropic Principle

What do scientists make of such anthropic coincidences? How do they explain them? Questions abound, among them: Is our universe uniquely "fine-tuned" to give rise to life and even human life? Must the universe have somehow known we were coming? Was the earth bound to be our home? Are anthropic coincidences geared toward us? Was the universe specifically made for us? Were we built in from the beginning?

Is the answer to these questions yes or no?

And if the answer is yes, can we then explain why?

As a matter of fact, scientists and philosophers have come up with a wide spectrum of explanations. They could all be grouped together under the heading "the anthropic principle." This principle is very broad and vague: it basically states only that we live in a universe that we can apparently live in. But from that point on, we can go in various directions. As Oxford University professor Nick Bostrom puts it, "A total of over thirty anthropic principles

---

[43] Hawking, *A Brief History of Time*, 7.

have been formulated and many of them have been defined several times over—in nonequivalent ways—by different authors, and sometimes even by the same authors on different occasions. Not surprisingly, the result has been some pretty wild confusion concerning what the whole thing is about."[44]

It is obvious that there are many reasons to be skeptical about simple understandings of anthropic coincidences, as if they must be there "for us." To put it differently in a simplistic way: If you win the lottery, were you meant to win? Thinking in terms of goals, ends, and purposes can easily be ridiculed with tricky questions. Are keys made for locks, or are locks made for keys? Were eggs made for chickens, or were chickens made for eggs? Is DNA there for us to propagate ourselves, or are we there for DNA so DNA can replicate itself? Not surprisingly, such examples make some anthropic principles suspect from the outset.

No wonder many people try to avoid any kind of anthropic principle. They may do so for various reasons or motives. One of them is that scientists don't like to go into theories that cannot be verified or falsified. Another reason is that they consider any anthropic principle unscientific, because it neither explains nor predicts any physical constants. Or they prefer to take on a professional-looking kind of skepticism. Or they consider the principle circular: the universe had to be this way, or else we would not be here to comment on it—which is rather trivial and begging the question. Or they counter that the physical environment of this universe did not have to get fine-tuned to life, but rather the other way around: life had to adapt to the physical environment. Or they argue that we cannot really know what is necessary for life to arise; it might well be, for instance, that life could be based on silicon

---

[44] Dick Bostrom, *Anthropic Bias: Observation Selection Effects in Science and Philosophy* (New York: Routledge, 2010), 6.

rather than carbon. Or they say that if our planet had been suitable for another kind of life, then that kind of life would have evolved here. Or, finally, they just don't want to open the door for any input from religion, for they consider science and religion archenemies. For any of these reasons, they hold on to their opinion that we were not "meant" to be here. To think differently causes discomfort and even hostility in much of the scientific community.

Nevertheless, there have been many attempts to come up with explanations as to why anthropic coincidences are the way they are. They try to explain the facts we are facing. In other words, anthropic *coincidences* are facts, while the anthropic *principle* is a collection of many speculative hypotheses for explaining those facts. Let's clarify this confusion without going into minute details. At the one end of the spectrum, there is a *weak* anthropic principle (WAP), and on the other end, a *strong* anthropic principle (SAP).

Let's start with the weak version, WAP. It does not deny the facts. For instance, it does not deny that Planet Earth has some unique features. Earth is not so close to the sun that we burn up, or so far away that we freeze solid. If it were much closer to the sun, it would be too hot to have liquid water; if much farther away, it would be too cold. If the earth were much smaller, it would not have sufficient gravity to retain an atmosphere. If it were much bigger, it would retain a lot of hydrogen in its atmosphere, which would be the wrong kind of atmosphere for life. Those facts are hard to deny. When it comes to the question "Has someone 'fine-tuned' the conditions here to make life possible," however, the WAP defenders say, "Not necessarily so."

The crucial question is how we explain that Planet Earth has some unique features. To explain such "coincidences," the WAP falls back on the concept of randomness and the statistical law of large numbers—which is basically a matter of hit-and-miss or trial and error. A common analogy is rolling dice. Dice have no memory of the past and no foresight (unless they are loaded). If you don't believe it, test it in the casino. The more you roll the dice, we say, the better your chances are of winning sooner or later. The same with the lottery: you have a better chance of winning the lottery if you buy more tickets and buy them more often.

The WAP applies this idea to planets. The more planets there are — according to the statistical law of large numbers (LLN) — the better the chances are that there will be a planet with the conditions we find on Planet Earth. It's like a winning ticket: coincidental, rather than fine-tuned. There are presumably a vast number of planets in the universe. Some planets are hot, some cold, some big, and some small. They undoubtedly span a wide range of physical and chemical conditions. It seems inevitable that some of them would happen to have the right conditions for life. This may also explain the enormous time span of our universe. The odds of a planet where life is possible and where life did emerge become better and better when we increase the number of years dramatically. That explains, some say, why the earth has to be so old, and the universe has to be even older.

Now, the idea of the WAP is that the argument can be used not just about planets but also about universes. Suppose that there are a huge number of universes. (There is much more to this discussion, which we will ignore here.) Let's just mention one representative of this approach: Stephen Hawking. He explains "fine-tuning" as randomness in disguise by postulating some "basic stuff" underlying universe formation. His explanation is worded by Michael Augros as follows:

> This fertile nothing spews out universes quite randomly, and so the fundamental constants in them must represent nearly all possible quantitative variations. Most of them, presumably, do not allow life. But some of them inevitably

will. And the fact that we happen to live in a life-friendly one is neither a matter of design nor a wild coincidence.[45]

How do we assess the weak version of the anthropic principle? Well, the WAP seems to be an explanation that fits within a scientific framework, for it works with concepts, such as randomness, that have a scientific allure to them. On the other hand, the WAP cannot be scientifically proven itself—it remains a mere hypothesis. To avoid acceptance of its rival, the SAP, the WAP had to come up with a random explanation: spewing universes. Perhaps the best assessment about the WAP is that it gives scientists an easy alibi to avoid going into its much stronger version, the SAP.

What does the *strong* version of the anthropic principle claim? Unlike the WAP, the SAP doesn't seek refuge in randomness. It does consider anthropic coincidences as coincidental, but only seemingly so. It acknowledges that the odds against a universe like ours appear to be enormous and cannot be ignored. Unlike the WAP, the SAP attempts to explain why such improbable events occurred, rather than just state that improbable events must have occurred in the universe to let us be here. And besides, it acknowledges that all of these far-fetched coincidences happened together! It remains still amazing that there is such a close match between what physics determines and biology requires. Could that be mere coincidence, too?

Stephen Barr makes the case against the WAP even stronger by stressing that, if the order of the universe were the result of mere chance, as the WAP asserts, we still have not explained how this orderliness could be so perfect that the laws of nature apply to everything anywhere in the universe, without any exceptions. His point is that we find *perfect* order and lawfulness in our universe:

[45] Michael Augros, *Who Designed the Designer? A Rediscovered Path to God's Existence* (San Francisco: Ignatius Press, 2015), 14.

Among all the logically possible universes, ones that have the perfection of order and lawfulness that ours displays are highly exceptional, just as among all possible rocks, a perfect gem that has absolutely no flaws in it is almost infinitely unlikely. Why doesn't our universe exhibit occasional departures from its regularities — the regularities we call the laws of physics — just as gemstones have occasional departures from their regularities? No answer to this is possible.[46]

The weak version of the anthropic principle doesn't have the answer to this observation, for it cannot explain the perfect order and lawfulness in our universe that we discovered in the previous chapters. As a matter of fact, the laws of nature appear to have a highly interesting and extraordinarily beautiful mathematical structure. No longer can we take things for granted as matters of mere fact.

Even someone like Stephen Hawking had to acknowledge, "The laws of science, as we know them at present, contain many fundamental numbers.... The remarkable fact is that the values of these numbers seem to have been very finely adjusted to make possible the development of life."[47] The theoretical physicist Freeman Dyson put it this way, "As we look out into the universe and identify the many accidents of physics and astronomy that have worked together to our benefit, it almost seems as if the universe must in some sense have known that we were coming."[48]

---

[46] Stephen M. Barr, "Anthropic Coincidences," *First Things* (June 2001).

[47] Hawking, *A Brief History of Time*, 125.

[48] Freeman J. Dyson, *Disturbing the Universe* (New York: Basic, 1979), 250. See also "Energy in the Universe," *Scientific American* 224 (1971): 50.

So, it seems to me that the SAP has better answers than the WAP. If some object that the SAP is not testable in a scientific sense, they must concede that the WAP isn't either. The weak version with its idea of "spewing universes" is perhaps even harder to test than the strong version. As Stephen Barr noted, "It seems that to abolish one unobservable God, it takes an infinite number of unobservable substitutes"[49]—such as the plurality of universes in Hawking's view. The SAP at least accepts that there is fine-tuning, but this does not mean, of course, that the fine-tuning is done by God, who has an end or goal in mind for the universe and for humanity, since that requires consciousness and foresight.

Would that last viewpoint be possible? Let's find out.

Does Science Have the Last Word?

Not one of the explanations we have seen so far for the existence of anthropic coincidences is very convincing. Each one is, in essence, a nearly desperate move to avoid the only sensible alternative—that the universe was designed by a Creator God who had us in mind. According to that scenario, we were predestined to emerge, and the earth was predestined to become our home. Behind it all, there must be a divine plan, the work of a Divine Mind.

For some scientists, it is still hard to make this step. The astronomer Fred Hoyle, for instance, seems to have come halfway. Once an outspoken atheist, he had the courage to exclaim, "A common sense interpretation of the facts suggests that a superintellect has monkeyed with physics, as well as with chemistry and biology."[50] The legendary Albert Einstein came to a similar aware-

---

[49] Barr, *Modern Physics and Ancient Faith*, 157.
[50] Fred Hoyle, "The Universe: Some Past and Present Reflections," *Engineering and Science* (November 1981): 12.

ness: "Everyone who is seriously involved in the pursuit of science becomes convinced that a Spirit is manifest in the laws of the universe—a Spirit vastly superior to that of man."[51] As the physicist John Wheeler put it, "A life-giving factor lies at the center of the whole machinery and design of the world."[52] And as the physicist Paul Davies said, "There must be an unchanging rational ground in which the logical, orderly nature of the universe is rooted."[53]

In spite of his agnostic stance, Stephen Hawking was intellectually honest enough to ask a rather profound philosophical question that may open the gate to the SAP:

> What is it that breathes fire into the equations and makes a universe for them to describe? The usual scientific approach of constructing a mathematical model cannot answer the questions of why there should be a universe for the model to describe. Why does the universe go to all the bother of existing?[54]

All of this makes the quest for a Designer not only highly relevant but even inescapable. I see no other solution than that the order of the universe is the result of a design that was implemented—actually conceived, invented, and decreed—by a Divine Designer who had a divine plan. In other words, our universe is a

[51] *Albert Einstein's Letters to and from Children* (Amherst, NY: Prometheus Books, 2002), 127–129. Also quoted in Max Jammer, *Einstein and Religion: Physics and Theology* (Princeton, NJ: Princeton University Press, 1999), 93.
[52] John A. Wheeler, foreword to John D. Barrow and Frank J. Tipler, *The Anthropic Cosmological Principle* (Oxford, UK: Clarendon Press, 1986), vii.
[53] Paul Davies, "What Happened before the Big Bang?" in Russell Stannard, ed., *God for the 21st Century* (Philadelphia: Templeton Press, 2000), 10–12.
[54] Hawking, *A Brief History of Time*, 174.

work of cosmic engineering, with God as the Engineer and Law-giver. As the book of Wisdom says about Him, "You have arranged all things by measure and number and weight" (11:20).

A divine plan requires a Divine Mind with perfect knowledge and perfect foresight. You probably wonder how God can have foresight of things that haven't happened yet. The answer is that God is in no way a temporal being, but rather the Creator of time, with complete and equal access to all its contents. But if God exists entirely outside of time—in a kind of eternal present to which all that occurs in time is equally accessible—He would indeed be able to comprehend all of history, the past and the present as well as the future, just as though they were now occurring. He is a God not merely of the past but also of the present and the future. He is the eternal "I AM," the One "who is and who was and who is to come" (Exod. 3:14; Rev. 1:4).

Even things that we call "random" from a human or natural point of view may still very well be included in God's eternal, providential plan. God knows infallibly and from eternity what is going to happen to the universe. This is not some kind of secret knowledge. St. Thomas Aquinas argues that God's infallible knowl-edge of our future is not a secret predictive power, because there is no such thing as the future for God. Rather, in His transcendent eternity, which is outside the flow of time, all events from any time are present to Him in one eternal now. But seen from our "tem-porary now" perspective, we are still waiting for things to happen.

Of course, this is hard for us to understand. Perhaps a simple analogy will help us to see that there is no contradiction between God's divine plan and foreknowledge, on the one hand, and what is happening in the universe based on physical constants and laws of nature, on the other hand. If you were watching a video of events in your past life, you may get the impression that these actions were, in fact, predetermined and preordained, and yet you know that they were partly determined by preceding events and partly freely decided upon by human beings. Well, God in His eternity is like someone watching an eternal "video" of what is, was, and will be in the universe without taking anything away from the role that laws of nature and physical constants play in this process. Of course, this is just an analogy, and therefore inadequate.

God is not the direct cause of my decisions—I am—but He is the indirect cause that lets me be the cause of my decisions. Seen in this light, our human freedom need not be in conflict with an all-knowing God. God is the complete and Primary Cause of a free act, whereas the human person is its complete secondary cause. God's willing that I decide as I do does not make my decision God's. God's willing does not take away from me the operations of my free will, or the actions founded upon them; they remain mine.

There is another problem that some people might have with the idea of a divine plan: it is the fact that the plan seems to be so much out of proportion. We easily feel lost in this universe with its enormous size and its immense age. We seem almost to be drowning in its vastness. The late astrophysicist Carl Sagan is known to have called our planet "a speck of dust in the universe."[55] Blaise Pascal had a similar feeling when he confessed, "The eternal silence of the infinite spaces frightens me."[56] Yet, it is this speck of dust that we call home. If this planet is a speck of dust, we must be even tinier specks of dust.

This raises at least two serious questions: (1) Why does the universe have to be so *old?* and (2) Why does the universe have to be so *big?* Well, in answer to the first question: chemicals had to be formed in stars and then be released when those stars exploded

---

[55] Carl Sagan, *Pale Blue Dot: A Vision of the Human Future in Space* (New York: Ballantine Books, 1997).

[56] Blaise Pascal, *Pensées*, trans. T. S. Eliot (New York: E. P. Dutton, 1958), 61.

as supernovas. The whole lifetime of a star is typically billions of years. Living creatures cannot appear until at least some stars have had time to explode and release the elements needed for biochemistry. As to the second question: because of its expansion, the size of the universe is directly related to its age. It has grown to a size of at least fifteen billion years. If the universe were any smaller, it would not have lasted long enough for life to emerge. Seen in this light, the vastness of the divine plan makes more sense than ever.

Of course, there remains another, perhaps more serious, problem: if the universe was made toward some end, then something with enough consciousness and foresight to create so precise a universe has to exist. Many scientists don't want to take this step, because the idea of a Creator God—a God whom science cannot observe and therefore cannot prove exists—goes against their scientific instinct.

True, God is certainly not an object we can observe, as is done in science. As we have seen, however, there are very compelling philosophical reasons to conclude that only a Primary Cause, God, whom we cannot possibly see and observe, can make all secondary causes, which we do observe, possible. Secondary causes are causes *within* nature, while God, the Primary Cause, is the cause *of* nature. God is not one of the entities we can observe. We cannot haul God into our laboratories for further investigation and experimentation, for He is not a secondary cause, but God makes secondary causes possible. The question "Does God exist?" is not like "Do neutrinos exist?" God cannot be "trapped" by some kind of ingenious experiment. He is, in the words of the Nicene Creed, the Creator of all that is seen and unseen, of all that is visible and invisible—without being visible and seen Himself. Because God is omnipresent, he only *seems* to be nowhere.

The "case for God" can be made even stronger when we realize that scientists could not do their work without the silent, often forgotten assumption that God is the very foundation of their work. This may sound enigmatic at first, but the reason for this is rather obvious once you give it serious thought. Here is why.

God is the ultimate source of the order as well as the intelligibility of the universe. Let's start with the *order* of the universe. That the universe is orderly is not something that is immediately obvious or intuitive; nor is it something that science has discovered after an extensive search. There is no way that scientists could prove that there is order in this universe. That scientific evidence can refute a scientific hypothesis—which is called falsification—is possible only if there is order in this universe; without order, falsifying evidence could not exist, for it would just be taken as an exception or a random occurrence. Falsification and verification are not possible if there is no order in the universe. In other words, order in the universe is not a discovery but an assumption—actually a religious belief—that must come first before science can even begin.

Something similar could be said about the *intelligibility* of the universe. That the universe is intelligible and comprehensible is not something that is immediately obvious or intuitive; nor is it something that science has discovered after an extensive search. There is no way we could prove that the universe is intelligible and comprehensible. If people say that the universe cannot be comprehended, we have nothing to prove they are wrong, other than telling them to keep searching. In other words, intelligibility is not a discovery but an assumption—actually a religious belief—that must come first before science can even begin. Scientists assume that the world can be understood and taken as intelligible—otherwise there is no reason to pursue science.

Yet, the order and the intelligibility of the universe should not be taken for granted. Albert Einstein, in utter amazement, wrote in one of his letters, "But surely, a priori, one should expect the world to be chaotic, not to be grasped by thought in any way."[57] Albert Einstein also used to say that the most incomprehensible thing about the universe is that it is comprehensible.[58]

So, the question is "Where do the assumptions of order and intelligibility come from, if they are not intuitive?" Certainly not from science. In fact, they come from the Judeo-Christian belief in a Creator God. At the very core of the Judeo-Christian view is the view that the universe is the creation of a Divine Mind—a rational Intellect that is capable of being rationally interrogated by all human beings, including scientists. How could nature be intelligible if it were not created by an intelligent Creator? How could there be order in this world if there were no orderly Creator?

There is more and more evidence that the Judeo-Christian faith in a Creator God with a divine plan furnished the conceptual framework in which science came to fruition. Therefore, it would be dangerous, actually suicidal, for scientists to cut off the Judeo-Christian roots they came from. Arguably, science could hardly have gotten off the ground if the Judeo-Christian religion had not created the right framework for its arrival. Without the religious idea that the universe is a rational and intelligent creation endowed with order and intelligibility, science would most likely never have emerged.

One could even make the case that, without God, scientists have no reason to trust their scientific reasoning. Denying the existence of God in essence eats away the very foundation of science. In light of this, one might argue that all scientists keep living off Judeo-Christian capital, whether they like to admit it or not. They borrow from the Judeo-Christian faith what they themselves cannot provide.

[57] Albert Einstein, letter to M. Solovine.
[58] One may say "the eternal mystery of the world is its comprehensibility." Albert Einstein, "Physics and Reality," *Journal of the Franklin Institute* (March 1936), reprinted in *Out of My Later Years* (New York: Gramercy Publishing, 1993).

Not surprisingly, the history of science shows us that it is through their belief in a Creator God that many scientists found reason to investigate nature. Science helped them to "read" God's divine plan for the universe, so to speak. Here are some witnesses.

- *Astronomer Johannes Kepler*: The chief aim of all investigations of the external world should be to discover the rational order and harmony which has been imposed on it by God.[59]
- *Berkeley Nobel laureate Charles Townes*: For successful science of the type we know, we must have faith that the universe is governed by reliable laws and, further, that these laws can be discovered by human inquiry.[60]
- *Nobel laureate in medicine and physiology Joseph Murray*: The more we learn about creation—the way it emerged—it just adds to the glory of God.[61]
- *Physicist Paul Davies*: People take it for granted that the physical world is both ordered and intelligible.... Science can proceed only if the scientist adopts an essentially theological worldview.[62]
- *Pope John Paul II*: It is the one and the same God who establishes and guarantees the intelligibility and reasonableness of the natural order of things upon which scientists confidently depend.[63]

---

[59] Johannes Kepler, *De fundamentis astrologiae certioribus*, Thesis 20.

[60] Charles H. Townes, "Logic and Uncertainties in Science and Religion," Pontifical Academy of Sciences, *Scripta Varia* 99 (2001): 300.

[61] In an interview for the *National Catholic Register*, December 1–7, 1996.

[62] Paul Davies, "Physics and the Mind of God: The Templeton Prize Address," *First Things* (August 1995).

[63] John Paul II, Encyclical *Fides et Ratio* (September 14, 1988), no. 34.

But there is at least one more enigma left: How could human minds possibly "read" the Divine Mind behind God's divine plan for the universe?

The book of Genesis gives us the solution to this enigma. It tells us that the *human* mind was created as a reflection of the *Divine* Mind: "Then God said, 'Let us make man in our image, after our likeness'" (1:26, RSVCE). Because it is a dim reflection of the Divine Mind, the human mind has at least basic access to the Divine Mind and its divine plan.

As we found out in the previous chapters, the physical order we observe in this universe appears to be amazingly consistent. It follows natural laws that can be unified into higher-level theories with great elegance and harmony—and perhaps someday even into a single unified theory.

It is a consistency that must perplex us, for how is it possible that reality can be grasped by the human mind? Not only is the rationality of the human mind consistent, but so is the universe itself. The conformity between the rationality of the human mind and the "rationality" of the universe may be a riddle for nonbelievers, but for religious believers, it is actually "a match made in Heaven." The mystery we have here is the fact that the rationality present in our minds matches the "rationality" we find in the divine plan of the universe.

Even scientists uphold the conviction—consciously or subconsciously—that the universe has an elegant, intelligible, and discoverable underlying mathematical and physical structure, a structure that is accessible to the human mind. This made the late astrophysicist Sir James Jeans exclaim, "The universe begins to look more like a great thought than a great machine."[64] The

---

[64] James Jeans, *The Mysterious Universe* (Cambridge: Cambridge University Press, 1930), chap. 5.

"great thought" that Jeans speaks of is a thought not of the human mind but of the Divine Mind. Now that it has been found that the laws of nature as discovered by science form a single magnificent edifice of great subtlety, harmony, and beauty, Stephen Barr finds reason to declare, "The question of a Cosmic Designer seems no longer irrelevant, but inescapable."[65]

I must admit that the assumption of a Creator or Divine Designer does not explain *why* the universe is the way it is, but it is still the best explanation of the fact *that* the universe is this way. I do not know of a better explanation.

Where does this power of the human mind come from?

The answer is that both the "rationality" of the universe and our capacity to understand it have the same ultimate source. Only the rationality of the Divine Mind with its divine plan can explain that the world is an objective and orderly entity that can be investigated by the human mind, because the mind, too, is an orderly and objective product of the same rational and consistent Creator.

Fr. Georges Lemaître spoke about the God of the Big Bang as the "One Who gave us the mind to understand him and to recognize a glimpse of his glory in our universe which he has so wonderfully adjusted to the mental power with which he has endowed us."[66]

It's time to come to a conclusion. The idea of a divine plan goes much further than the claims of the WAP and the SAP. The WAP, and probably the SAP to a lesser degree, claim it goes too far. Do they have a point? I don't think so. The notion of spewing universes is as unobservable as the idea of a Creator God with a divine plan. We have shown, however, that the idea of a Creator

---

[65] Stephen Barr, "Retelling the Story of Science," *First Things* (March 2003): 16–25.

[66] Georges Lemaître, *The Primeval Atom* (New York: D. Van Nostrand, 1950), 55.

In the Beginning

God is a necessary foundation for the scientific enterprise. It is also a necessary condition for the existence of secondary causes. As St. Thomas Aquinas put it, "God is to all things the cause of being."[67]

Besides, no matter how improbable this universe and our Planet Earth are, it is still amazing that all of these far-fetched coincidences happened together and that there is such a close match between what physics determines and biology requires. It has become harder and harder to maintain this is mere coincidence, and not the result of a divine plan. For that reason, *unique* might be more adequate than *improbable* to characterize the status of Planet Earth.

Ultimately, we cannot avoid the question "Why is there something rather than nothing?" Even if all physical constants and laws of nature can someday be derived from one grand unified theory, we can't say that what determined all physical constants and laws of nature is this unified theory—and not God. For we are still left with the question as to where that grand unified theory came from, given the fact that the universe could very well have been based on other theories or frameworks. We would still be left, as Paul Davies puts it, "with the mystery why the universe has the nature it does, or why there is a universe at all."[68] The answer to that question leads us almost inevitably to a divine plan in the Divine Mind of a Creator—which would mean we were predestined to be here.

---

[67] Thomas Aquinas, *Summa Contra Gentiles*, 2, 46.
[68] Paul Davies, *God and the New Physics* (New York: Simon and Schuster, 1983), 42.

5

The Evolution of the Universe

After the Big Bang, the universe was "populated" with structures such as galaxies, stars, and planets. How did those structures come along? Specifically, how did the Milky Way galaxy and our solar system come about?

The Milky Way

Once the universe cooled sufficiently, its energy was converted into various atomic and subatomic particles, including protons, neutrons, and electrons. Giant clouds of these primordial elements then coalesced through gravitation to form galaxies, which are systems of stars, star remnants, interstellar gas, dust, and dark matter, held together by gravitation.[69] In general, astronomical bodies were formed by gas and dust that condensed through the force of gravity.

A combination of observation and theory suggests that the first galaxies were formed about a billion years after the Big Bang, and since then, larger structures have been forming, such as galaxy clusters and superclusters. Populations of stars have been in the

---

[69] L. S. Sparke and J. S. Gallagher III, *Galaxies in the Universe: An Introduction* (Cambridge: Cambridge University Press, 2000), i.

process of aging and evolving, so that distant galaxies (which are observed as they were in the early universe) appear very different from nearby galaxies (observed in a more recent state).

Galaxies range in size from dwarfs, with just a few hundred million stars, to giants, with one hundred trillion stars each orbiting its galaxy's center of mass. The majority of galaxies are gravitationally organized into groups, clusters, and superclusters.

One of these galaxies is what is widely known as the Milky Way, part of a local group of galaxies that form part of the Virgo supercluster. The Milky Way is home to Planet Earth, the birthplace of humanity.

The Milky Way is a spiral galaxy with a diameter between 100,000 and 180,000 light-years. If you could look down on it from above, you would see a central bulge surrounded by four large spiral arms that wrap around it. There are probably at least a hundred billion stars in the Milky Way.[70] This number is not fixed, however, because the Milky Way is constantly losing stars through explosions of supernovas and producing new stars—about seven per year. Our galaxy has been described as an "exceptionally quiet" spiral galaxy.[71] Its magnetic field is relatively weak yet strong enough to prevent the collapse of its spiral structure. Its disk is dense enough to sustain the spiral structure.

The Milky Way has grown by merging with other galaxies through time. It formed over billions of years in a process that involved interactions between smaller galaxies, in particular the

---

[70] A. Cassan et al., "One or More Bound Planets per Milky Way Star from Microlensing Observations," *Nature* 481 (January 11, 2012): 167–169.

[71] F. Hammer, Mathieu Puech, L. Chemin, and Hector Flores, "The Milky Way, an Exceptionally Quiet Galaxy: Implications for the Formation of Spiral Galaxies," *Astrophysical Journal* 662 (June 2007): 322-334.

gradual capture of many stars from nearby dwarf galaxies. In the last ten billion years, however, it has undergone no mergers with large galaxies, which is unusual among similar spiral galaxies; its neighbor the Andromeda galaxy appears to have a more typical history shaped by more recent mergers with relatively large galaxies.

All of this may appear like a chaotic, random sequence of events. Yet the outcome is surprisingly structured—not just a random assortment of stars, like the grains of sand on a beach. Instead, our galaxy is a highly evolved entity with an elegant structure that shows both order and complexity. Its surprisingly beautiful shape is so common among galaxies that the universe almost seems to delight in building them. The end product is especially remarkable in light of what is believed to be the starting point of our galaxy: nebulous blobs of gas.[72]

The gas blobs that evolved into the Milky Way consisted merely of hydrogen and helium (and a bit of lithium), which were created in the Big Bang. All other elements were created by the stars. Stars are prodigious factories of elements. So, the presence of certain elements gives us an indication about the age of planets and stars. Since iron, for instance, is produced very gradually inside the "furnace" of a star, lack of iron shows that a system (including its stars), didn't have enough time to produce iron, and thus must date from a time before elements such as iron became abundant.

Our Solar System

Our solar system—composed of the sun (a star) and all the planets around it—is part of the Milky Way, about twenty-six thousand light-years from its center. This raises the question of the origin

[72] Cristina Chiappini, "The Formation and Evolution of the Milky Way," *American Scientist* (November–December 2001): 506–515.

In the Beginning

of our solar system—that is, the gravitationally bound system of the sun and the objects that orbit the sun.

Basically, scientists have learned that several billion years ago, our solar system was nothing but a cloud of cold gas and dust particles swirling through largely empty space. This cloud of gas and dust was disturbed—perhaps by the explosion of a nearby star (a supernova)—and started to collapse as gravity pulled everything together, forming a huge spinning disk. As it spun, the disk separated into rings. The center of the disk grew to become the sun, and the particles in the outer rings turned into large fiery balls of gas and molten liquid that cooled and condensed to take on solid form. About 4.5 billion years ago—at 4:17 p.m. on the scale of a day, and on September 3 on the scale of a year—they began to turn into the planets in our solar system.

The solar system consists of the sun, the 8 official planets, at least 3 dwarf planets, more than 130 satellites of the planets (moons), a large number of small bodies (comets and asteroids), and the interplanetary medium. This classification has become rather blurry, as there are several moons larger than Pluto and two larger than Mercury; there are many small moons that probably started out as asteroids and were only later captured by a planet; and comets sometimes fizzle out and become indistinguishable from asteroids. But no matter what, here is the punch line: they all follow a predictable path according to the same basic laws of nature—no real surprises left.

The four smaller inner planets, Mercury, Venus, Earth, and Mars, are terrestrial planets, primarily composed of rock and metal. The four outer planets are substantially more massive than the terrestrials. The two largest, Jupiter and Saturn, are gas giants, being composed mainly of hydrogen and helium; the two outermost planets, Uranus and Neptune, are ice giants, composed mostly of water, ammonia, and methane.

The existence of rocky planets depends on the presence of nearby gas giants, with a mass of at least ten times Earth's mass. These giants protect the rocky planets from taking too many destructive hits from asteroids and comets by pulling those objects away. Thus, they provide a protective shield.[73]

The smallest known star is only eighty-three times the mass of Jupiter. The largest known planet is likely less than ten times Jupiter's mass. The size of these planets really matters. Larger planets would get crushed by their own gravity and become stars. Smaller planets would not be big enough for gravity to pull them into a spherical shape; they would be asteroids with no atmosphere. In other words, to exist, the size of a planet must be within a certain limited range.

The orderly arrangement and patterns of planets (close to being circular) and of the solar system (with all its orbits almost in the same plane, looking like a giant pinwheel) emerged in a natural way out of a swirling, cooling, and condensing cloud of dust and gas. It was gravitation that caused the chaotically swirling gas and dust that filled the universe after the Big Bang to condense into stars and planets. The same force made those stars and planets settle into orderly arrangements as in our solar system. It's randomness curtailed by gravitation.

To stay intact, the solar system itself had to be ejected outward in the opposite direction from the galactic center. Interestingly, astronomers have found that a galactic ring of both minimum stellar density and minimum gas density exists right at the edge of our solar system.[74]

All eight planets have elliptical orbits. These orbits all lie approximately in the same plane, and all the planets go around in

[73] Hugh Ross, *Improbable Planet* (Grand Rapids, MI: Baker Books, 2016), 45.

[74] Ross, *Improbable Planet*, 40.

their orbits in the same direction. Gravity is the force that determines the shape of the orbits. If the strength of gravity were not based on the inverse square of the distance ($1/d^2$), the orbits of the planets in the solar system would be much more complicated than ellipses and would in most cases not form closed curves at all. As a consequence, the earth and the other planets would not be able to remain in stable orbits around the sun.

What are the elements we find on these planets? The 92 naturally occurring elements in the universe were not all there at the very beginning of the universe but were manufactured in the fires of the Big Bang, in the interiors of stars, and in the explosions of stars called supernovas. During the Big Bang, the first nuclei consisted of lone neutrons or lone protons. The simplest form of hydrogen (H), for instance, has a lone proton in its nucleus (hydrogen-1). A lone proton can be fused with a lone neutron to make hydrogen-2 (deuterium). Other nuclei can be made by fusing them together into larger nuclei. Once deuterium is made, deuterium nuclei can combine by fusion to make the nucleus of helium, which has two protons. These steps happen very readily. Then the nuclei get "dressed" in a cloud of electrons to make a complete atom.

The next step in the sequence from the smallest element to the largest element is a bit more complicated. When two helium nuclei collide in the interior of a star, they cannot fuse permanently, but they remain stuck together for an extremely short period. In that tiny sliver of time, a third helium nucleus comes along and hits the other two. Three helium-4 nuclei, as it happens, do have enough sticking power to fuse together permanently. When they do so, they form a nucleus called carbon-12.

A sequence of more and more such fusions eventually led to the element uranium (U), with 92 protons and 92 electrons (and its isotopes with various numbers of neutrons). That's how Planet Earth eventually acquired its 92 "naturally occurring" elements, such as silicon (Si) and oxygen (O). These elements can then be incorporated into numerous kinds of minerals, such as quartz (SiO_2).

Looking back in time, we cannot but acknowledge that it was indeed a very protracted process leading from the Big Bang to the formation of the solar system—and eventually to the formation of Planet Earth. It makes us realize that the "order in the heavens" comes from an underlying order in the laws of nature.

Seeming Chaos

At first sight, it may seem hard to relate the previous processes in the developing universe to God. They appear to be rather haphazard and random. Was all of this really "done" to make Planet Earth our home? All we have seen so far makes a rather chaotic impression—as if "ruled" by chaos.

When people think of chaos, they usually think of a mess. They see the universe as a "mess" of swirling galaxies and wildly orbiting planets, and they see the early earth as a "mess" of colliding tectonic plates and imploding meteors. As we found out, however, that impression is based on a misconception. Galaxies, planets, and even meteors don't follow chaotic paths. They go where the laws of nature, especially the law of gravity, take them. Their paths may appear messy, but they are not. They are inherently part of a world of law and order.

Yet some scientists still like to talk about "chaos" in the universe as part of their antireligious agenda. The British chemist Peter Atkins, for instance, tells us, "We are children of chaos."[75] That almost sounds as if he's honoring some kind of deity called "Chaos," which leaves us with the deity of chaos versus the God of

---

[75] Peter Atkins, *The Second Law* (Scientific American Library, 1984), 200.

order. Contrast this with what the prophet Isaiah says about God: "[He] formed the earth and made it (he established it; he did not create it a chaos, he formed it to be inhabited!)" (45:18, RSVCE).

True, recently, some scientists have developed a legitimate interest in "chaotic" systems. What they mean is that some natural systems, such as the weather, can be described only by nonlinear mathematical equations with such complex solutions that we cannot exactly predict what the system will do in the near future. This isn't really chaos in the sense of a "mess," however. Scientists are, in fact, working to puzzle out the very order behind these seemingly chaotic phenomena.

On the other hand, although there is no mere chaos in the universe, there is still *randomness*. Randomness is, in essence, a statistical or stochastic concept that plays an important role in science, and therefore in the description of the universe. Let's take a simple example of rolling dice, which is basically a stochastic process. Randomness is involved because the outcome is independent of what the one who rolls the dice would like to happen; it is also independent of previous and future rolls. The key is that randomness has no memory and no foresight. For example, we cannot know the outcome of a dice roll before it occurs, nor can we explain how it turns out the way it does; there are too many variables and causes acting together, such as the speed at which we throw the dice and the way the dice bounce when they first hit the table.

Yet, in the aggregate, we are able to make predictions to a certain extent, but that has to be done in terms of probabilities. For example, if someone rolls two dice, the outcome of any particular roll is unpredictable, but a sum of 7 will likely occur twice as often as a sum of 4. In this view, randomness is a measure of uncertainty of what the final outcome will be. According to the statistical law of large numbers, the frequency of different outcomes over a large number of events is somewhat predictable. The science of probabilities is based on very orderly probability distributions; it helps us to put a meaningful numerical value on things about which we do not have enough detailed knowledge. Whereas a single random event may not be predictable, the aggregate behavior of random events is. But it's certainly not chaos.

This is the kind of randomness that science deals with. Randomness has become a key word in the vocabulary of most scientists. But don't confuse randomness with chaos (in the sense of a mess). Randomness is not chaos. Let's look at three examples of randomness in science.

The first example concerns a meteor hitting the earth. That event may look like chaos, but the paths of the earth and the meteor are well defined, far from chaotic. What is random about the event is that the two paths happen to intersect and coincide at one point, which could be called coincidental. Something like that also happens when the path of a falling stone happens to cross the path of where I am walking. But this doesn't mean that there is no order behind random occurrences such as these.

Randomness is always restrained by order. It can work only within a framework of order that determines its boundaries. A previous order must exist before any random event can occur. Randomness itself could never create the order found in the universe—as blindness cannot create sight. If there were no preexisting order, there could be no randomness, because randomness needs the order of preexisting causes coming together to produce unexpected results—which is usually called "coincidence."

When many independent causal chains intersect with one another, there is statistical randomness. In the same vein, everything on Earth is the product of a long chain of mostly improbable events. Blaise Pascal once remarked, "If the nose of Cleopatra had been shorter, the whole face of the earth would have been changed."[76]

---

[76] Blaise Pascal, *Thoughts* (1623), 8, 29.

Another example of randomness is radioactivity and radioactive decay. Randomness has become prominent in physics, particularly in quantum physics, which is inherently probabilistic in nature. The laws of physics would permit one to calculate only the relative probabilities of various future outcomes, called quantum indeterminacy. This became already very clear in the case of radioactive decay, where the laws of physics allow one to assign only a probability for it to happen within a certain period. We discussed this phenomenon for isotopes and radioactive dating techniques. We cannot predict or determine when a particular atom will disintegrate—even though we can register the individual decay event on a Geiger counter. The decay of a particular atom is therefore considered unpredictable, but in the aggregate, we can make predictions based on probabilities. That's how we can predict that 50 percent will decay in a half-life period. It's again falling back on the statistical law of large numbers.

A third example of randomness is the occurrence of mutations in DNA, which holds the genetic code of living beings. Changes in DNA can be caused by radiation and certain chemicals. Mutations are considered random in the sense of being spontaneous—they just pop up somewhere in the DNA, and we don't know when they will occur. They are also unpredictable: we do not know at which exact location in the DNA mutations will take place. Mutations are also random in the sense of being arbitrary, because mutations change their target indiscriminately, which means they have no preference for the changes they might generate—good and bad results alike, so to speak. Mutations are also random in the sense of being aimless, because they occur without any connection to the immediate or future needs of the organism—in that sense, they can also be considered "shortsighted." There is no physical mechanism that detects which mutations would be beneficial and then causes those mutations to occur—in that sense, they certainly lack foresight. Fortunately, however, organisms have developed mechanisms to repair certain effects caused by mutations.[77]

The randomness of mutations has caused the most discussion among scientists. The general opinion is that randomness is "blind"; it has no "favorites," no "memory," and no "foresight." In 2005, thirty-eight Nobel laureates issued a statement defending evolutionary theory, stating, "Evolution is understood to be the result of an unguided and unplanned process of random variation and natural selection."[78]

Randomness and Providence

We seem to have reached a stalemate here. How can the idea of randomness ever be reconciled with the idea that God is behind what happens in the universe? We have here a potential conflict between randomness in nature and divine providence, between a chaotic outcome and a divine plan.

What is in charge: Is it randomness or providence?

This becomes even more pressing when we realize that randomness has no foresight, whereas divine providence does. No wonder many scientists reject the idea that God built a house for us. The most they might agree to is that randomness built a house for us.

But that's not the end of the discussion. As a matter of fact, randomness and providence can go together. To explain this, we can take advantage of an important distinction that Stephen Barr makes. When we speak of randomness in science, we are talking in statistical terms, in the sense of how things in this universe are

---

[77] Cindy Chang, "DNA Repair," *Encyclopaedia Britannica*, https://www.britannica.com/science/DNA-repair.

[78] Open letter sent to the Kansas State Board of Education by thirty-eight Nobel laureates in 2005.

related to *one another* (for example, we exist thanks to our parents). But when we speak of a divine plan, we are talking about how things in this universe are related to God—instead of one another (we exist thanks to God).[79] Put in different terms, randomness concerns the relationship between secondary causes, whereas divine providence is about the relationship of secondary causes to the First Cause.

When we look at randomness and chance in relation to God, we get a picture very different from what randomness seems to suggest. Seen in relation to God, there is no randomness, but divine providence instead. The development of the universe, of the earth, and of humanity is not one immense lottery, but a guided process seen from God's point of view. The *Catechism* says about divine providence: "The universe was created 'in a state of journeying' (*in statu viae*) toward an ultimate perfection yet to be attained, to which God has destined it. We call 'divine providence' the dispositions by which God guides his creation toward this perfection" (302). As the Nobel laureate in chemistry Christian Anfinsen put it, "We must admit that there exists an incomprehensible power or force with limitless foresight and knowledge that started the whole universe going in the first place."[80]

Not only can things that are random be part of divine providence, but also science can neither prove nor disprove God's providence. Stephen Barr is right when he says:

> No measurement, observation, or mathematical analysis can test whether or not God planned a development like a genetic mutation. What apparatus would one employ?

[79] Stephen M. Barr, "Chance, by Design," *First Things* (December 2012): 27.

[80] As cited in Henry Margenau and Roy A. Varghese, eds., *Cosmos, Bios, Theos* (La Salle, IL: Open Court Publishing, 1997), 139.

Being "unplanned by God" is simply not a concept that fits within empirical science. Being "statistically random," on the other hand, is, because it can be tested for.[81]

The doctrine of divine providence does not tell us what the divine plan is—neither in general nor in detail. All it tells us is that the universe has unfolded exactly as known and willed by God from all eternity. As the *Catechism* tells us, "God cares for all, from the least things to the great events of the world and its history" (303). God's providence is not just some general oversight of the universe. Rather, God is the direct cause of every detail of the universe.

In other words, a divine plan still allows for randomness, but when scientists elevate randomness or chance to an all-inclusive principle of the universe, they are stepping outside the boundaries and limitations of science. As Stephen Barr rightly remarks, "When biologists start making statements about processes being unsupervised, undirected, unguided, and unplanned, they are not speaking scientifically."[82] They capitalize the term *chance* and make it a kind of deity that replaces God as the First Cause of the universe. Scientists as scientists are not qualified to make such judgments. At best, they are showing their antireligious sentiments. The question as to whether the universe has a meaning, destination, or purpose takes us into the domains of philosophy and religion.

The fact that there is some randomness in this universe doesn't mean that the universe is completely ruled by chance. Randomness doesn't rule out that the universe was designed by God, who had us in mind. Seen this way, we are still entitled to say that our

---

[81] Barr, "Chance, by Design," 26.
[82] Ibid.

universe is rather "fine-tuned" instead of accidental, coincidental, or random.

The *Catechism* puts all of this together as follows: The world "is not the product of any necessity whatever, nor of blind fate or chance" (295). In his first homily as pontiff, in 2005, Pope Benedict XVI insisted: "We are not some casual and meaningless product of evolution. Each of us is the result of a thought of God. Each of us is willed, each of us is loved, each of us is necessary."[83] This is and remains true, even in a world seemingly stripped by science. That's why randomness and divine providence are not complete opposites. Scientists tend to lose sight of the larger picture when they focus on the scientific details. That's what Alexander Pope noticed about the atheists in the mid-eighteenth century:

> See Nature in some partial narrow shape,
> And let the Author of the Whole escape.[84]

Let's bring this discussion about randomness to a conclusion. First, randomness has nothing to do with chaos. Second, randomness is a scientific concept and therefore is confined to the domain of science. Third, randomness still leaves ample space for divine providence.

---

[83] Benedict XVI, Homily at the Mass for the inauguration of his pontificate (April 24, 2005).
[84] Alexander Pope, *The Dunciad*, bk. 4, 455–456.

6

How the Earth Developed

In the middle of numerous immense processes taking place in the universe, there appears at one point a rather insignificant newcomer: Planet Earth. Here, preparations are being made for what most people consider very significant developments: the rise of life and of human beings. But before that is possible, quite a bit of groundwork has to be done.

Planet Earth

Evidence from radiometric dating indicates that the earth is about 4.54 billion years old.[85] Earth began to form—at 4:17 p.m. on a day scale, or September 3 on a year scale—from the same cloud of gas (mostly hydrogen and helium) and interstellar dust that formed our sun, the rest of the solar system, and even our galaxy. In fact, Earth is still forming and cooling from the galactic implosion that created the other stars and planetary systems in our galaxy.

The earth went through a period of catastrophic and intense formation during its earliest beginnings. After the formation of

---

[85] G. Brent Dalrymple, "The Age of the Earth in the Twentieth Century: A Problem (Mostly) Solved," *Special Publications, Geological Society of London* 190, no. 1 (2001): 205–221.

the planets in our solar system, tens of thousands of asteroids and comets pounded Mars, Earth, the moon, Venus, and Mercury.[86] During that time, 90 percent of the moon's craters were formed. It is estimated that Earth alone received an average of two hundred tons of bombardment material per square yard. These bombardments delivered water to our planet, for asteroids can be more than 10 percent water and comets can be up to 85 percent water. They also loaded our planet with heavy metals and altered the tilt of Earth's rotational axis by as much as 10 degrees.[87]

Condensing water vapor, augmented by ice delivered by comets, accumulated in the atmosphere, cooled the molten exterior of our planet to form a solid crust, and produced the oceans. Thus, by some 4 billion years ago, Earth had become a planet with an ocean and an atmosphere (different from our atmosphere today, though). This period of Earth's formation is referred to as the Precambrian period—making up roughly seven-eighths of Earth's history (from 4.6 billion to 500 million years ago, which runs from the beginning of September to the middle of December on the scale of a year). It marks a very long period prior to the Cambrian period, when hard-shelled creatures began to appear in abundance.

The atmosphere of early Earth is not well understood. Most geologists believe it was composed primarily of nitrogen, carbon dioxide, water vapor, and other relatively inert gases and was lacking in free oxygen (even modern volcanic gases contain no

[86] Ariel D. Anbar, Kevin J. Zahnle, Gail L. Arnold, and Stephen J. Mojzsis, "Extraterrestrial Iridium, Sediment Accumulation and the Habitability of the Early Earth's Surface," *Journal of Geophysical Research: Planets* 106 (February 2001): 3219–3236.

[87] William F. Bottke, Richard J. Walker, James M. D. Day, David Nesvorny, Linda Elkins-Tanton, "Stochastic Late Accretion to Earth, the moon, and Mars," *Science* 330 (December 2010), 1527–1530.

oxygen). In the Precambrian atmosphere, carbon dioxide concentration was hundreds of times higher than now (as it still is in the atmospheres of Venus and Mars). The high content of greenhouse gases in early Earth's atmosphere must have prevented heat-radiation loss into space. This caused a high surface temperature.

Combine this with the sun's comparatively low luminosity during the Precambrian period. Luminosity is the total amount of energy emitted per unit of time by an astronomical object. Since its birth 4.5 billion years ago, the sun's luminosity has very gradually increased by about 30 percent.[88] This is an inevitable evolution because the sun has been burning up the hydrogen in its core. The helium "ashes" left behind are denser than hydrogen, so the hydrogen-helium mix in the sun's core is very slowly becoming denser, thus raising the pressure. This causes the nuclear reactions to run a little hotter, which makes the sun brighter.

Earth is at such a distance from the sun that the temperature of its surface during different periods has changed, but within a rather small range. If the luminosity of the sun had been 1.5 to 2 times greater, Earth would have been like Venus, with a dense atmosphere of carbon dioxide and steam. On the other hand, if the sun's luminosity were lower, Earth could have frozen, like Mars.

Over the last four billion years, as the sun's luminosity has been increasing, Earth's carbon dioxide mass has been decreasing due to the reduction in gas release from Earth's upper mantle. When the sun's luminosity was still low and Earth's carbon dioxide content still high, however, the atmosphere kept Earth's surface temperature within the limits needed for the development of primitive life.

[88] Ignasi Ribas, "The Sun and Stars as the Primary Energy Input in Planetary Atmospheres," *Proceedings of the International Astronomical Union Symposium* 264 (February 2010): 3–18.

In the Beginning

About this period, Hugh Ross notes that "if Earth had retained its primordial inventories of water and carbon, its ocean would have been so deep that continents would never have arisen, and its atmosphere so thick that lungs would have been incapable of functioning."[89]

But things on Earth began to change. Short-lived isotopes generated the heat that blasted away Earth's primordial light gases and water that would otherwise have left Earth with an atmosphere far too thick and an ocean far too deep for life to be possible. Earth nowadays has about twelve hundred times less carbon-based atmospheric gas and about five hundred times less liquid water than planets with the same mass as Earth and orbiting at the same distance from a sun-like star.[90] The atmosphere we know—with oxygen (21 percent), nitrogen (78 percent), and carbon dioxide (0.04 percent)—arose mainly as a result of volcanic activity and release of gases from the interior part of Earth.

Then there is the earth's strong and enduring magnetic field, which protects life from deadly cosmic and solar radiation and prevents Earth's atmosphere from being sputtered away by solar particles. Certain long-lived isotopes, such as iron, uranium, and thorium, play a critical role in establishing this magnetic field. For a rocky planet like the earth to maintain a sufficiently strong and enduring magnetic field, it must have a particular internal composition—for instance, a liquid iron outer core surrounding a solid iron inner core.[91]

---

[89] Ross, *Improbable Planet*, 166.

[90] Linda T. Elkins-Tanton and Sara Seager, "Ranges of Atmospheric Mass and Composition of Super-Earth Exoplanets," *Astrophysical Journal* 685 (October 2008): 1237–1246.

[91] Jorge I. Zuluaga, Sebastian Bustamante, Pablo A. Cuartas, and Jaime H. Hoyos, "The Influence of Thermal Evolution in the Magnetic Protection of Terrestrial Planets," *Astrophysical Journal* 770 (June 2013): 23.

The composition of the interior part of the earth and its crust had also quite an impact on the minerals we find on the planet. Eight elements account for most of the key components of minerals, due to the elements' abundance in the earth's crust, where most of the minerals are derived. These eight elements, composing more than 98 percent of the crust by weight, are, in order of decreasing abundance: oxygen (O), silicon (Si), aluminum (Al), iron (Fe), magnesium (Mg), calcium (Ca), sodium (Na), and potassium (K). Oxygen and silicon are by far the two most important—oxygen composes 47 percent of the crust by weight, and silicon accounts for 28 percent.

Another peculiarity of Planet Earth is its abundance of heavier metals, which means that it was formed closer to the center of the Milky Way, where the abundance of elements heavier than helium is near peak value.[92] This also explains why Earth's crust is so abundant in (nonradioactive) aluminum.

Earth is the largest and densest of the inner planets in our solar system. Its liquid hydrosphere is unique among the terrestrial planets, and it is the only planet where plate tectonics has been observed. Today, Earth's atmosphere is radically different from those of the other planets, having been altered by the presence of life to contain 21 percent free oxygen.

The Continents

Another peculiarity of planet Earth is its strong *plate tectonic* activity—the movement of parts of Earth's crust—which creates continents. Earth's outer shell is divided into several plates that glide over the mantle, which is the rocky inner layer above the core. The plates act like a hard and rigid shell compared with Earth's

---

[92] Ross, *Improbable Planet*, 37

mantle. This strong outer layer is called the lithosphere, which is one hundred kilometers thick.

Today, Earth's crust comes in two flavors. *Oceanic* crust is mafic, made of dark, iron-and-magnesium-rich rocks such as basalt that come directly from the melting of Earth's mantle (the layer just beneath the crust). *Continental* crust, on the other hand, formed from the melting of mafic rocks, is felsic, made of lighter-colored rocks such as granite, which is rich in silicon and aluminum.

Early Earth must have sported a mafic crust. It's an open question when felsic rocks first started forming. Figuring out when felsic continental crust formed would mean identifying a start date for plate tectonics. That's because subduction zones—places where tectonic plates crash into each other and oceanic crust slides beneath the continental crust—serve as the primary factories for felsic rocks. Subduction zones bring water down into the crust, which lowers the melting point of rock by disrupting the bonds in the minerals within the rock. This leads to the formation of the felsic rocks that make up continents.

Earth began as a water world. Tectonic activity first produced volcanic islands, many of which eroded away. Ongoing tectonics generated granitic rocks, which are less dense than basaltic rocks and which floated above them to make continents. This caused Earth's rotation to slow down from two to four hours per day to its current rate of twenty-four hours per day. This change meant that its clouds could reflect more sunlight and thus keep Earth's surface cooler to compensate for the warming effects of the brightening sun.[93]

No one has ever been able to show exactly when plate tectonics began. Several studies had traced the beginning of plate tectonics back to around 3 billion years ago, but new research suggests that

[93] Ross, *Improbable Planet*, 135

this dynamic started earlier, some 3.5 billion years ago—that is, only about a billion years after the formation of the planet. Part of the problem is that there aren't many rocks left on Earth that date back billions of years to when the planet had just formed. Of rocks that are that old, most have been altered by weathering and chemical processes over the eons.

What starts and sustains plate tectonics? We don't know exactly, but we do know that the driving force behind plate tectonics is convection in the mantle—a movement caused by the tendency of hotter and therefore less dense material to rise, and colder, denser material to sink under the influence of gravity. Therefore, hot material near Earth's core rises, and colder mantle rock sinks. We also know that, without sufficient quantities of heavy, long-lived radioactive isotopes such as thorium and uranium, a planet would lack the interior flow that is crucial for launching and sustaining plate tectonics. Relative to the element magnesium, thorium is 610 times more abundant in Earth's crust than it is in the rest of the Milky Way. Uranium is 340 times more abundant. This made the mantle hotter and lowered its viscosity, which led to larger, faster currents in the mantle some 3.5 billion years ago. Heavy bombardment of Earth generated variations in thickness and density of the crust, which allowed slabs to slide over and under each other.

Sometimes tectonic movements break up continents, but at other times they occasionally bring them together to form supercontinents, single landmasses. The forming of supercontinents and their breaking up into continents appears to have been cyclical through Earth's history. One supercontinent formed some 2 billion years ago and broke up about 1.5 billion years ago. Another supercontinent is thought to have formed about 1 billion years ago and to have embodied most or all of Earth's continents, and then was broken up into eight continents around 600 million years ago. Later, these eight continents reassembled into another supercontinent called

Pangaea (or Pangea). It assembled from earlier continental units approximately 335 million years ago, and it began to break apart about 175 million years ago.[94] When breaking apart, the continents migrated into their current positions and alignment. These plates still move relative to each other, typically at rates of 5 to 10 centimeters per year, and they interact along their boundaries, where they converge, diverge, or slip past one another. Pangaea was the most recent supercontinent to have existed and the first to be reconstructed by geologists.

Whatever Earth looked like before plate tectonics, these powerful tectonic forces define the world as it is today—with seven or eight "major" plates: African, Antarctic, Eurasian, North American, South American, Pacific, and Indo-Australian. The diving and crashing of tectonic plates—which became better known as "continental drift"—not only created the continents we know and live on today but also recycles minerals and nutrients through Earth's system.

The Earth-Moon System

How did the moon arise? The elliptical orbit of the moon may reflect its origin. Current evidence suggests that the moon-forming event occurred about one hundred million years after the formation of Earth,[95] probably caused by a collision of Earth with a protoplanet about the size of Mars. As said earlier, giant impacts must have been rather common in the early solar system. The debris

[94] J. J. W. Rogers and M. Santosh, *Continents and Supercontinents* (Oxford: Oxford University Press, 2004), 146.

[95] W. F. Bottke, D. Vokrouhlický, S. Marchi, T. Swindle, E. R. D. Scott, J. R. Weirich, and H. Levison, "Dating the Moon-Forming Impact Event with Asteroidal Meteorites," *Science* 348 (April 2015): 321–323.

from this collision bonded together to form the moon. The impact released a lot of energy, which would have melted the outer shell of Earth, and thus formed a magma ocean.[96] This collision delivered iron and other critical elements to Earth's core and mantle, as well as long-lasting isotopes, the heat from which now drives most of Earth's tectonic activity and volcanism.[97] As a matter of fact, rocks from the moon had the same isotopic signature—the ratio of stable isotopes to unstable radioactive isotopes of particular elements—as rocks from Earth. They differed, however, from almost all other bodies in the solar system.[98]

The collision produced a moon with sufficient mass to stabilize the angle of Earth's rotational axis, protecting the planet from rapid and extreme climatic variations. The moon also gradually slowed Earth's rotation rate. Since the moon and Earth are held together by gravity and Earth is much more massive than the moon, the moon orbits Earth. Each orbit takes 27.3 days. The moon also rotates, or spins on an internal axis once every 27.3 days. The rotation and revolution take the same amount of time, so it makes one rotation per revolution. However, because Earth is moving in its orbit around the sun at the same time, it takes slightly longer for the moon to show the same phase to Earth, which is about 29.5 days. The Earth-moon system also revolves around the sun, taking 365.25 days (or a year) to complete one orbit.

[96] W. Brian Tonks and H. Jay Melosh, "Magma Ocean Formation Due to Giant Impacts," *Journal of Geophysical Research* 98 (1993): 5319–5333.

[97] Peter D. Ward and Donald Brownlee, *Rare Earth: Why Complex Life Is Uncommon in the Universe* (New York: Copernicus, 2000): 191–234.

[98] U. Wiechert, A.N. Halliday, D.-C. Lee, G. A. Snyder, L. A. Taylor, and D. Rumble, "Oxygen Isotopes and the Moon-Forming Giant Impact," *Science* 294, no. 12 (October 2001): 345–348.

In the Beginning

Earth's moon is fifty times larger than any other moon in the solar system and orbits more closely than any other large satellite yet discovered. This is what gives Earth a stable rotational-axis tilt. If the moon were a bit more massive, it would pull the tilt out of balance. If it were less massive, it would have taken longer than 4.5 billion years to slow Earth's rotation rate from three to four hours per day at the time of the moon-forming event to twenty-four hours per day. A more rapid rate would have caused greater temperature extremes and less evenly distributed rainfall.[99] Besides, without the presence and configuration of Earth-moon system, the orbits of Venus and Mercury would be affected in such a way that the inner solar system would be out of "balance."

As a side note: the moon, together with the sun, would give Earth's future inhabitants a "clock" to calculate days, months, seasons, and years—important for a "home." We use the Gregorian calendar, which replaced the much older Julian calendar in 1582. Using the best calculations available, Julius Caesar had extended the year to 365.25 days long and added at least 67 days to that year, bringing January 1, 45 B.C. in line with the start of winter. And finally, to make the year as self-regulating as possible, he had built in a leap day to recur every four years after February 24 to make up for the quarter day.

The Romans, however, had not been able to calculate the length of the solar year with complete accuracy. So, by the sixteenth century, the calendar year was ten days behind—that is, it was ten days behind the sun. Around 1560, the Council of Trent had required some reform of the liturgical calendar, and this would need some corrections of Julius Caesar's calculations. These corrections were made by Jesuit astronomers in what came to be called the Gregorian calendar.

---

[99] Hilke E. Schlichting, Paul H. Warren, and Qing-Zhu Yin, "The Last Stages of Terrestrial Planet Formation: Dynamical Friction and the Late Veneer," *Astrophysical Journal* 752 (June 2012): 8.

The reason for the new calendar was that the average length of the year in the Julian calendar was too long, as it treated each year as 365 days, 6 hours in length, whereas calculations showed that the actual mean length of a year is slightly less—365 days, 5 hours, 49 minutes. So, the years in the Julian calendar were about 11 minutes too long, thus losing about three days every 400 years. To correct for this, the new calendar would eliminate three leap days every 400 years. Yet everyone should be happy with the Gregorian calendar—it's almost perfect. Some researchers, however, have suggested that another tweak will be due by the year 4000, when the Gregorian system will have gone off by about a day. But for the next 1,980 New Years or so, we should be able to make any plans we'd like.

The Role of Chance in God's Plan

A striking observation in the history of Planet Earth is that what looks like a series of coincidences worked together in an amazing process of harmony. Had one or several of these coincidences been different, the end result would not have been favorable for what is coming next on Planet Earth.

This takes us back again to the question how God's providence can be harmonized with so many coincidences during the formation of Planet Earth. Does not one exclude the other? Not according to St. Thomas Aquinas. On the one hand, Aquinas does not dismiss or deny the role of chance in God's creation. According to William E. Carroll, Aquinas "argues that God causes chance and random events to be the chance and random events which they are, just as he causes the free acts of human beings to be free acts."[100] On the other hand, Aquinas does not deny or dismiss divine providence

---

[100] William Carroll, "Evolution, Creation, and the Catholic Church," lecture at Williams College, October 19, 2006.

either. He once said, "Whoever believes that everything is a matter of chance, does not believe that God exists."[101]

This latter conviction was also shared by St. Padre Pio. He was apt to say in various ways that it is God who arranges the coincidences. He asked a man who claimed such-and-such event had happened by chance: "And who, do you suppose, arranged the chances?" Science has no answer to this question—not even the answer "nobody did." Science always creates a fragmented reality that only religion can piece together. Therefore, anything that seems to be random from a scientific point of view may very well be included in God's eternal plan.

If this is true, there is no longer a conflict per se between the role of chance and the working of divine providence. As the book of Proverbs says, "The lot is cast into the lap, but the decision is wholly from the LORD" (16:33, RSVCE).

The Bible even has several examples in which randomness and chance play a decisive role—for instance, by casting lots. The eleven Apostles used chance to have God elect a new twelfth Apostle to replace Judas Iscariot: "Then they gave lots to them, and the lot fell upon Matthias" (Acts 1:26). Priests were chosen by lot as it says about Zechariah: "According to the practice of the priestly service, he was chosen by lot to enter the sanctuary of the Lord to burn incense" (Luke 1:9). The Old Testament mentions another case: "Then Saul said, 'Cast the lot between me and my son Jonathan.' And Jonathan was taken" (1 Sam. 14:42, RSVCE).

We can learn from this that things that are random can very well be part of divine providence. Chance and providence do not exclude each other. There is a role for chance in God's plan for this universe.

---

[101] Aquinas, *De Symbolo Apostolorum*, 4, 33.

7

The Evolution of Life

It is on the "tiny" Planet Earth with its remarkable blend of life-friendly conditions that life arises in all its diversity, ranging from simple unicellular life to the complex life of human beings.

The Dawn of Life

All life requires the presence of molecules that are much more complex than simple minerals—molecules such as DNA, RNA, and proteins. Nowadays proteins require DNA to form, and DNA needs proteins to form, so how could these have formed without each other? The answer may be RNA, which can store information like DNA, but can also help create both DNA and proteins—the so-called RNA-world hypothesis.[102] The idea is that the very first RNA molecules formed from three chemicals: a sugar (called a ribose); a phosphate group; and a base, which is a ring-shaped molecule of carbon, nitrogen, oxygen, and hydrogen atoms. It was shown that a chemistry based on cyanide (CN) could make two

---

[102] Shelley Copley, Eric Smith, and Harold J. Morowitz, "The Origin of the RNA World: Co-Evolution of Genes and Metabolism," *Bioorganic Chemistry* 35, no. 6 (December 2007): 430–443.

of the four units in RNA and many amino acids.[103] According to this hypothesis, it was only later that DNA and proteins, which are more efficient, succeeded this "RNA world" to make room for a "DNA world."

How could complex molecules develop from much simpler elements and minerals? Years ago, in the famous Miller-Urey experiment reported in 1953, it was shown that electric sparks can generate amino acids and sugars from an atmosphere loaded with water (H_2O), methane (CH_4), ammonia (NH_3), and hydrogen (H_2). This suggests that lightning might have helped create the key building blocks of life on Earth in its early days. Scientists have suggested since then that volcanic clouds in the early atmosphere were filled with lightning and might have held methane, ammonia, and hydrogen as well. Electromagnetic forces seem to have the capacity to make chemicals clump together into small and then larger molecules.

All so-called organic molecules, essential to life, are carbon based. The chemical element carbon (C) has the capacity to form very long chains of interconnecting C-C bonds, which allows carbon to form an almost infinite number of new organic compounds. This is why all life on Earth is carbon based, making carbon the "favorite" building block of the living world on Planet Earth.

But even if we can explain the presence of the most elemental organic molecules, there is still a long way to go from complex molecules to even the simplest organism. The simplest organism known today is the bacterium *Pelagibacter ubique*, one of the smallest and most abundant organisms in the ocean. It consists of

---

[103] Bhavesh H. Patel, Claudia Percivalle, Dougal J. Ritson, Colm. D. Duffy, and John D. Sutherland, "Common Origins of RNA, Protein and Lipid Precursors in a Cyanosulfidic Protometabolism," *Nature Chemistry* 7, no. 4 (2015): 301–307.

1,354 genes, which are made up of DNA, and it obtains its energy without oxygen. The simplest organism that produces oxygen is the bacterium *Prochlorococcus marinus*, which has 1,884 genes. Do we have any indication that such simple organisms existed in the early history of Planet Earth?

When life first emerged on Earth, it was in the form of unicellular organisms.[104] There is a fairly solid record of bacterial life throughout the Precambrian period from 3.5 billion years on. How do we know we are dealing with life here? One of the indications is the presence of what is called a "life signature" associated with life — a higher ratio of lighter versus heavier isotopes. This is considered a "life signature," because remains of life contain higher ratios of lighter isotopes (carbon-12 over carbon-13, nitrogen-14 over nitrogen-15, sulphur-32 over sulphur-34). This has led to an early origin-of-life date: 3.8 billion years ago — as old as the oldest sedimentary rocks now known.[105] Earth is about 4.5 billion years old, but the oldest rocks still in existence date back to just 4 billion years ago. Apparently and surprisingly, not long after that rock record begins, tantalizing evidence of life emerges.

So, it seems very likely that life arose on Earth as soon as solid rocks and liquid water began to exist with some stability on the planet's surface. Since Earth's heavy bombardment peaked at about 3.9 billion years ago, life's origin had a very narrow time window of "only" 0.1 billion years for starting the long road from complex molecules to simple organisms. This happened at approximately

---

[104] Yanan Shen, Roger Buick, and Donald Canfield, "Isotopic Evidence for Microbial Sulphate Reduction in the Early Archaean Era," *Nature* 410 (March 2001): 77–81.

[105] Craig E. Manning, Stephen J. Mojzsis, and T. Mark Harrison, "Geology, Age, and Origin of Supracrustal Rocks in Akilia, West Greenland," *American Journal of Science* 306 (May 2006): 303–366.

5:08 p.m. on a day's scale, or in the middle of September on a year's scale.

Somehow, life managed to emerge on Earth very rapidly soon after the oceans had condensed on the planet's surface, some 4.5 billion years ago. What this means, according to many scientists, is that life may not be such a difficult process to start once we have the right conditions and ingredients, properly prepared for over a long time since the Big Bang. In that sense, paleontologist Niles Eldredge said it right: "In the very oldest rocks that stand a chance of showing signs of life, we find those signs."[106]

Besides, we should realize that the number of genes is not directly correlated with the complexity of an organism. Humans do not have many more genes than a tiny roundworm needs to manufacture its utter "simplicity." And humans have only 300 unique genes not found in mice. No wonder the president of a bio corporation said about this surprising finding, "This tells me genes can't possibly explain all of what makes us what we are."[107] Or as Francis Collins, the former leader of the U.S. contingent to the Human Genome Project, put it: "One surprise is just how little of the genome is actually used to code for protein.... The total amount of DNA used by those genes to code for protein adds up to a measly 1.5 percent of the total.... Our complexity must arise not from the number of separate instruction packets, but from the way they are utilized."[108]

But how and where could life emerge? When we think of life, we often have in mind plants and animals that need oxygen to survive.

---

[106] Niles Eldredge, *The Triumph of Evolution and the Failure of Creationism* (New York: W.H. Freeman, 2000), 35–36.

[107] Craig Venter, president of Celera Genomics, in *The San Francisco Chronicle*, February 13, 2001.

[108] Francis S. Collins, *The Language of God: A Scientist Presents Evidence for Belief* (New York: Free Press, 2006), 124–125.

But oxygen was not present in the early atmosphere. Oxygen only gradually became an important part of the atmosphere, mainly thanks to a process called photosynthesis, which lets green plants produce food with energy from the sun. That was not possible, however, during the early stages of life. Instead, chemosynthesis was the way to go. Chemosynthesis is the biological conversion of one or more carbon-containing molecules (usually carbon dioxide or methane) and other nutrients into organic matter using the oxidation of simple compounds such as hydrogen gas, hydrogen sulfide, or methane as a source of energy, rather than sunlight, as done in photosynthesis.

So now the question is how chemosynthesis could come along. There are several theories, but the "deep-sea vent" theory offers one of the most attractive scenarios. It suggests that life began at deep-sea hydrothermal vents, which spew out mineral-rich water. It is thought that the chemicals and energy found at these vents could have provided perfect conditions for primitive life to form. In 1977, scientists discovered biological communities unexpectedly living around seafloor hydrothermal vents, far from sunlight and thriving on a chemical soup rich in hydrogen, carbon dioxide, and sulfur, which were spewing from the geysers.[109] There are numerous species of organisms currently living immediately around deep-sea vents, suggesting that this is indeed a possible scenario. These vents host complex communities fueled by chemicals dissolved in the vent fluids. These bacteria are basically "rock-eating"; for instance, to produce organic material, they "eat" sulfur compounds, particularly hydrogen sulfide, a chemical highly toxic to most known organisms.

[109] P. Lonsdale, "Clustering of Suspension-Feeding Macrobenthos Near Abyssal Hydrothermal Vents at Oceanic Spreading Centers," *Deep-Sea Research*. 24, no. 9 (1977): 857–863.

In the Beginning

Inspired by these findings, scientists proposed that hydrothermal vents provided an ideal environment with all the ingredients needed for microbial life to emerge on early Earth. This is an appealing hypothesis because of the abundance of methane, ammonia, and sulfate present in hydrothermal-vent regions, a condition that was not provided by the earth's primitive atmosphere.

A major limitation of this hypothesis is the lack of stability of organic molecules at high temperatures, but some have suggested that life would have originated at some distance from the zones of highest temperature. Further evidence supports this scenario. Research on 355 proteins shared by all bacterial lineages allowed researchers to reconstruct the first shared relative from which all life today has descended, which suggests that it lived in an oxygen-free, hydrothermal environment, similar to the microbes that cluster around deep-sea vents today.[110]

Obviously, there are still many essential steps missing on the road from simple molecules to complex organisms — even organisms as "simple" as bacteria — but they are so technical that we will skip them here. Let's assume for now that rather primitive "rock-eating" bacteria evolved before any other life-forms, some 3.5 billion years ago during the Precambrian period. They obtained energy by chemically modifying minerals made of iron or sulfur found in rock. They were metabolizing "rock food" without the presence of oxygen. Many such bacteria are still alive today. An important group of such organisms at the beginning of life were the sulfate-eating bacteria.[111] They feed on sulfates, which are highly soluble and toxic in soluble form, and metabolize them

---

[110] Madeline C. Weiss, et al., "The Physiology and Habitat of the Last Universal Common Ancestor," *Nature Microbiology* 1 (September 2016): 1–8.

[111] Ross, *Improbable Planet*, 167–168.

into insoluble sulfides. Thus, they transform their environment from toxic habitats to habitats that are safe for more advanced life-forms.

It is due to this long history of progressively more advanced life-forms that the mineral inventory in Earth's crust grew from 250 minerals to its present level of 4,300 distinct minerals. This is another reason why Planet Earth is far from ordinary, especially in its abundance of elements and diversity of minerals.

Another important step in the evolution of life was the appearance of 3.7-billion-year-old colonies of blue-green algae (cyanobacteria), which form layered structures called stromatolites.[112] These new communities were anchored to the seashore close to sun and water, so they probably weren't eating minerals found in rock. Instead, they must have harvested energy through photosynthesis, suggesting that such microbes had switched from using rocks for their energy to using light energy. So, they were the first microbes to produce oxygen by photosynthesis. They convert sunlight into energy and produce oxygen as a waste product. Back then, the earth's atmosphere didn't have free oxygen in it, as it does today. It was locked up in water molecules or bonded to iron in minerals.

But things were changing: there were now organisms producing oxygen. The effect was not immediately noticeable. That was for two reasons. As these new organisms produced free oxygen, iron would bond with it, and thus, the environment could keep up with the production. Another reason was probably the presence of methane in the atmosphere. Early chemosynthetic organisms

---

[112] Allen P. Nutman, Vickie C. Bennett, Clark R. L. Friend, Martin J. Van Kranendonk, and Allan R. Chivas, "Rapid Emergence of Life Shown by Discovery of 3,700-Million-Year-Old Microbial Structures," *Nature* 537 (2016): 535–538.

likely produced methane, which is an important trap for molecular oxygen, since methane readily oxidizes to carbon dioxide and water in the presence of ultraviolet radiation. But once these oxygen "traps" had become saturated, they could no longer absorb the oxygen being produced. So, oxygen built up in the water and then in the air. To the other bacteria living in the ocean, oxygen was toxic; it killed countless species of microbes. It was the so-called Great Oxygenation Event, which caused a dramatic increase in atmospheric oxygen some 2.4 to 2.5 billion years ago. The mass extinction that followed was vast. But there was an important exception: some organisms could use that oxygen in their own metabolic processes. Combining oxygen with other molecules can release energy, a lot of it, and that energy is useful. It allowed these microscopic organisms to grow faster and breed faster. The predecessors of green plants had arrived.

On the other hand, the increased oxygen concentrations provided a new opportunity for biological diversification, as well as tremendous changes in the nature of chemical interactions between rocks, sand, clay, and other geological substrates and between Earth's air, oceans, and other surface waters.[113] Despite the natural recycling of organic matter, life had remained energy limited before the widespread availability of oxygen. This breakthrough in metabolic evolution greatly increased the free energy available to living organisms, with global environmental impacts.

The emergence of photosynthetic life modified Earth's geochemical cycles, which stimulated the production of granite and led to the first stabilization of continents, so that Planet Earth could transition from a water world to a world with both oceans

---

[113] Lynn Margulis and Dorion Sagan, "The Oxygen Holocaust" in *Microcosmos: Four Billion Years of Microbial Evolution* (University of California Press, 1986), 99.

and continents.[114] Without abundant photosynthetic life, plate tectonic activity on Earth would have been shut down, making the crust a stationary lid over everything beneath it. Without plate tectonics, Planet Earth would have become "permanently sterile," in the words of Hugh Ross.[115]

The Great Oxygenation Event (GOE) also triggered an explosive growth in the diversity of minerals, with many elements occurring in one or more oxidized forms near Earth's surface.[116] It is estimated that the GOE was directly responsible for more than 2,500 of about 4,500 minerals found on Earth today. Most of these new minerals were formed as hydrated and oxidized forms due to dynamic mantle and crust processes.[117] It should be noted, though, that time is of the essence here. Had the GOE occurred any earlier, the sun would have lacked the luminosity needed to prevent Earth from becoming a long-term ice ball.

So, we have a key event here. The sun was becoming brighter and brighter. At the time of life's origin, the sun was some 15 percent dimmer than it is today. This change required another set of compensations. One of them is silicate erosion. Silicates dominate the composition of growing continents and islands. Sea chemistry could not allow the formation of skeletons until a widespread continental erosion had caused a high abundance of phosphorus,

---

[114] Minik T. Rosing, Dennis K. Bird, Norman H. Sleep, William Glassley, and Francis Albarede, "The Rise of Continents—An Essay on the Geological Consequences of Photosynthesis," *Palaeogeography, Palaeoclimatology, Palaeoecology* 232 (March 2006): 99–113.

[115] Ross, *Improbable Planet*, 112.

[116] Dimitri A. Sverjensky and Namhey Lee, "The Great Oxidation Event and Mineral Diversification," *Elements* 6, no. 1 (2010): 31–36.

[117] Robert M. Hazen, "Evolution of Minerals," *Scientific American* (March 2010).

magnesium calcite, silica, and calcium carbonate.[118] As calcium silicate reacts with carbon dioxide, it produces calcium carbonate ($CaCO_3$) and sand (SiO_2). This promotes the removal of carbon dioxide from the atmosphere and adds carbonates and sand to the landmasses—which would open the door for crustaceans and vertebrates who depend on carbonates for their skeletons. This change in seawater chemistry occurred on a global scale, and so did the explosion of new organisms.

Apparently, after a start with very humble beginnings, life went through many stages. Each stage was a preparation for the next stage and set the conditions necessary for the next stage. For the first 2.8 billion years of life history, only unicellular life-forms existed. Multicellular organisms did not arise until 1 billion years ago, and animals no sooner than 600 million years ago. And finally, we see humanity appear—at 11:59 p.m. on the scale of a day, or December 29 on the scale of a year.

What Propelled Life?

As we found out earlier, life appeared on Earth relatively early and relatively quickly, after both the universe and the planet had been prepared. How could life emerge in such a short time? How could simple molecules assemble into complex molecules such as enzymes and DNA molecules, which are needed for life? There are at least three distinct explanations:

1. The statistical law of large numbers made it possible.
2. God's divine interventions took care of it.
3. The laws of nature made it possible.

[118] Shanan E. Peters and Robert R. Gaines, "Formation of the 'Great Uncomformity' as a Trigger for the Cambrian Explosion," *Nature* 484 (April 2012): 363–366.

Let's review each of them.

Explanation 1 is probably the best known. It is based on randomness and the statistical law of large numbers (LLN). It goes as follows: the origin of life may be a very improbable event on a single planet, but not if there are an enormous, or even infinite, number of planets; Planet Earth happened to be the "lucky" one.

The problem with an "infinite" number of planets is that we have no way of knowing that there is indeed an infinite number of planets. Again, this would replace an unobservable Creator with an unobservable infinitude of planets or even universes. Yet this explanation, or viewpoint, has received widespread attention because it has been heavily promoted by some scientists.

The statistical law of large numbers also comes in handy to explain how simple molecules could become assembled into complex molecules such as enzymes and DNA molecules, which are essential to life. The famous analogy used here is the one portraying an ape randomly hitting keys on a typewriter. According to this theory, an ape, hitting keys at random on a typewriter for an infinite amount of time will almost surely type a given text, such as the complete works of William Shakespeare. One of the earliest instances of the use of the ape theory is that of French mathematician Émile Borel in 1913.[119] Suppose the typewriter has fifty keys, and the word to be typed is *banana*. If the keys are pressed randomly and independently, it means that each key has an equal chance of being pressed. The chance that the first letter typed is *b* is $1/50$, and for all six letters $1/50^6$, which is less than one in 15 billion, but not zero, hence a possible outcome.

Could the ape ever produce one of Shakespeare's books this way? Evidently, that is very unlikely, unless we have an almost unlimited number of apes trying to achieve this. The physicist Arthur

---

[119] Émile Borel, "*Mécanique Statistique et Irréversibilité*," *J. Phys.* 5e série, 3 (1913): 189–196.

In the Beginning

Eddington expressed this idea as follows: "If an army of monkeys were strumming on typewriters they might write all the books in the British Museum."[120] Not only does this require many, many apes or trials; it would also take an enormous amount of imagination, as well as an enormous amount of time. It's almost a desperate search for randomness in order to avoid unwanted alternatives. Since life managed to emerge on Earth very rapidly rather soon after the oceans had condensed on the surface of the planet some 4.5 billion years ago, there must be a better scenario. So, it shouldn't surprise us, then, that most people reject this explanation and would like to find a better one.

Explanation 2 needs a longer discussion, because it has become the favorite in Christian circles, especially among Evangelical Protestants. It considers that the complexity of molecules, cells, and organisms cannot be explained by science and its theories alone. Therefore, it concludes that their complexity can be explained only by also invoking periodic divine interventions in addition to scientific explanations.

This idea is best known from the so-called intelligent design theory, or ID theory (held by authors such as Michael Behe, William Dembski, Michael Denton, Phillip E. Johnson, and Stephen C. Meyer). Its poster child is the bacterial flagellum (but they could have used other examples), which is an "outboard motor" that propels bacterial cells in various directions. Indeed, this "simple" flagellum does show quite an impressive case of complexity. Although not a friend of ID theory, the DNA expert Francis Collins describes the flagellum this way:

> The structure of the flagellum, which consists of about thirty different proteins, is really quite elegant. It includes

[120] Arthur Eddington, *The Nature of the Physical World: The Gifford Lectures* (New York: Macmillan, 1928), 72.

miniature versions of a base anchor, a drive shaft, and a universal joint. All of this drives a filament propeller. The whole arrangement is a nanotechnology engineering marvel. If any one of these thirty proteins is in-activated by genetic mutation, the whole apparatus will fail to work properly.[121]

This is certainly a masterpiece of design. ID theorists question whether a machinery of this level of complexity could ever have arisen step by step on the basis of random mutation and natural selection alone. So, they maintain that such complexity can be produced only by a Designer's intervention—an intelligent, supernatural cause—in addition to, or even instead of, the natural processes of mutation and natural selection.

Their main argument goes along these lines: How could any of these components have evolved by chance unless the other twenty-nine had developed at the same time? It does seem rather evident, they argue, that none of these parts could have evolved on a step-by-step basis until the entire structure had been assembled. Natural selection won't work, so the argument goes, until the entire structure has been assembled at once. Even a single mutation in one of the components makes the whole flagellum machinery collapse. Therefore, the ID theory claims that we need to invoke supernatural Designer interventions.

No wonder, considering features like these—including the emergence of life—it is very attractive to conclude that they are "irreducible" to natural, physical, or biological explanations.[122] On further investigation, however, the idea of intelligent design as an explanation for the "irreducible complexity" of life

---

[121] Francis Collins, *The Language of God* (New York: Free Press, 2006), 185.

[122] Michael Behe, *Darwin's Black Box* (New York: Free Press, 1998), 39.

faces some serious problems on both biological and philosophical grounds. On the biology front, it has been argued—especially by biologists in the neo-Darwinian camp—that "irreducible" complexity can, in fact, be reduced to step-by-step evolutionary processes of mutations and natural selection—without the need to invoke an additional Designer step. To show this, let us go back to the bacterial flagellum, the poster child of ID fans. Comparison of protein sequences from various bacteria has revealed that several components of the flagellum are related to an entirely different apparatus used by some bacteria to inject toxins into other bacteria—which in itself is an ideal target for natural selection. Most likely, the elements of this structure were duplicated in the "silent" parts of DNA and then recruited for a new use, subject to natural selection again. And in turn, this process could go on for other elements. In this case, selection just keeps working step by step, mutation by mutation, until the entire structure has been assembled.

Some neo-Darwinists use the following analogy to strengthen their case against the ID theory: once the keystone is placed in an arched stone bridge, the scaffolding can be removed; from then on, removal of any part may cause the bridge to collapse, but that doesn't mean the bridge wasn't the product of a gradual construction process.[123] So, when ID theorists raise the objection, "What good is half an eye," neo-Darwinists would reply, "Better than no eye at all."

This course of gradual construction can be exemplified by coagulation, also known as clotting. It is the process by which blood

---

[123] E.g., Niall Shanks, *God, the Devil, and Darwin: A Critique of Intelligent Design Theory* (Oxford: University Press, 2004), 185; A.G. Cairns-Smith, *Seven Clues to the Origin of Life: A Scientific Detective Story* (Cambridge: Cambridge University Press, 1986), 61.

changes from a liquid to a gel, forming a blood clot. So-called clotting factors activate each other in a complex cascade to form a blood clot. A partial clotting cascade can be found in fish: some of these proteins have a long history in the animal world, starting with a simple cascade—far better than none. But when it had to evolve from a low-pressure to a high-pressure cardiovascular system, our own blood-clotting mechanism had to gain the ability to stop leaks much more quickly—which required more proteins, most of them based on small changes in DNA duplicates. This complex mechanism did not come along all at once, but again, most likely step by step, mutation by mutation.

Another strategy of Neo-Darwinists is to attack the "intelligent" part of the ID theory. It is easy to bring in many examples of optimal designs—the best of what's available—that are not perfect designs, let alone intelligent designs. The fact that the oxygen-carrying protein in human blood, hemoglobin, has a three hundred times higher affinity for carbon monoxide (CO) than for oxygen can be detrimental with even very low concentrations of carbon monoxide. Such a feature does not seem to be a case of intelligent design; yet it is optimal, since natural selection did not have to deal much with high carbon monoxide levels in our evolutionary history.

Or consider that, during the course of their evolution, ruminants such as cows acquired very complicated stomachs (basically a set of four) and a very involved digestive routine. In this case, complexity doesn't seem a plus. As Philip Kitcher put it, "Their inner life would have been so much simpler had they been given the right digestive enzymes from the onset."[124] Or take the DNA sequence for an enzyme that makes ascorbic acid (vitamin C) in

---

[124] Philip Kitcher, *Abusing Science: The Case against Creationism* (Cambridge, MA: MIT Press, 1982), 139.

most animals. Many primates, including humans, have a defect in this DNA code. Consequently, these organisms must acquire vitamin C through the intake of food rich in vitamin C (which was abundant for arboreal primates anyway), but they did hold on to the original defective DNA code in the inactive section of their DNA.[125] It seems fair to ask why an intelligent Designer would first establish the vitamin C pathway and then abandon it.

So far, the arguments against the ID theory have come from the biological front. But what about the philosophical front? In general, the ID theory is not in line with the Catholic Church's position, especially not with Aquinas's philosophy and theology. How would Aquinas assess the ID theory? He would probably make the objection that the ID theory turns the universe from a comprehensive whole into a defective whole that is not causally complete in itself but requires instant and constant interventions to fill the gaps the Creator supposedly left in His creation. This kind of god often proves to be a fleeting illusion, for when the frontiers of science are being pushed back—and they usually are—this kind of god will be pushed back with them. As the theologian Dietrich Bonhoeffer famously put it, "We are to find God in what we know, not in what we don't know."[126]

The ID claim seems more of a desperate move similar to Isaac Newton's invocation of God's active intervention to reform the solar system periodically from increasing irregularities, to prevent heavenly bodies from falling in on one another. Today, we know that God doesn't have to make these interventions, because science can now explain them with the proper laws (which are God's

---

[125] I. B. Chatterjee, "Evolution and the Biosynthesis of Ascorbic Acid," *Science* 182 (1973): 1271–1272.

[126] Dietrich Bonhoeffer in a 1944 letter, in *Letters and Papers from Prison*, ed. Eberhard Bethge, trans. Reginald H. Fuller (New York: Touchstone, 1971), 311.

laws anyway). When Newton called on special interventions by the Creator in the working of the universe, the philosopher Gottfried Leibniz rightly quipped, "God Almighty needs to wind up his watch from time to time: otherwise it would cease to move. He had not, it seems, sufficient foresight to make it a perpetual motion."[127] This would mean, according to the philosopher Francis J. Beckwith, that "God creates everything ex nihilo and then returns now and again to tidy things up a bit when they seem to be going awry."[128]

Obviously, ID theorists basically ignore the distinction Aquinas makes between the Primary Cause and secondary causes and thus degrade the First Cause periodically to a secondary cause. They make the Primary Cause occasionally act like a secondary cause working inside the universe. Aquinas makes it very clear: "The same effect is not attributed to a natural cause and to divine power in such a way that it is partly done by God, and partly by the natural agent; rather, it is wholly done by both, according to a different way."[129] The great scholastic theologian Francisco Suárez put it this way: "God does not intervene directly in the natural order where secondary causes suffice to produce the intended effect."[130]

This insight gave the physicist Fr. Stanley Jaki good reason to praise Aquinas's idea that "the material realm is fully coherent, that is, it needs no special interventions from an outside factor,

---

[127] Gottfried Wilhelm Leibniz, "Mr. Leibniz's First Paper" (1715), in H. G. Alexander, ed., *The Leibniz-Clarke Correspondence* (Manchester, UK: Manchester University Press, 1998 [1956]), 11.

[128] Francis J. Beckwith, "How to Be an Anti-Intelligent Design Advocate," *University of St. Thomas Journal of Law and Public Policy*, 4, no. 1 (Fall 2009).

[129] *Summa contra Gentiles*, 3, 70, 8.

[130] Francisco Suárez, *De Opere Sex Dierum*, 2, 10, 13.

such as God, to keep it running."[131] Therefore, the ID theory may be not only bad biology, but also bad philosophy and bad theology. It degrades God to a glorified mechanic who steps in to arrange natural things when He feels like it. C. S. Lewis noted that a power outside the universe "could not show itself to us as one of the facts inside the universe."[132]

Because of the serious problems facing explanations 1 (mere randomness) and 2 (miraculous divine interventions), we may want to opt for explanation 3 to explain the emergence of life. This third position, which I think is by far the best, takes design very seriously, but not in the way the intelligent design theory does. Let me explain.

Randomness in the universe is harnessed by the laws and order of physical constants and the laws of nature. The universe was created with the design of law and order—I like to call it the "cosmic design." Even if we will never be able to derive these features from a unified physical theory, that very theory had to be implemented in the universe by God the Creator, who gave the universe His cosmic design.

It is thanks to this design that the impact of randomness is heavily curtailed—the outcome is "loaded." This means that the first simple molecules couldn't just react and combine with any other elements in a random way, as the typewriter analogy suggests.

Why not?

The world of chemistry is certainly not a game of making all kinds of random combinations of atoms or their reactions. Basically it's the physics behind it that determines which combinations

[131] Stanley Jaki, "The Biblical Basis of Western Science," *Crisis* 15, no. 9 (October 1997).
[132] C. S. Lewis, *Mere Christianity*, 3rd ed. (San Francisco: Harper, 2001), 21.

or reactions are possible. Helium and neon are the two lightest elements in a chemical group known as noble gases. These elements usually do not take part in chemical reactions because all the atoms in the group—helium, neon, argon, krypton, xenon, and radon—have outer electron shells that are fully filled, with neither any spare electrons to donate to other atoms nor any electron deficits that could be filled by electrons donated from other elements.

Electrons play an important role in all of this. Free radicals are atoms, molecules, or ions with unpaired electrons. With some exceptions, these unpaired electrons cause free radicals to be highly chemically reactive; they are prone to losing or picking up an electron, so that all electrons in the atom or molecule will be paired. Take, for instance, the free radical CO_3. In nature we find CO (carbon monoxide) and CO_2 (carbon dioxide) but no CO_3 (carbon trioxide). Yes, there is carbon trioxide, but it is so unstable that it degrades into CO_2 and H_2O with a lifetime much shorter than 1 minute.[133] To make a long story short, chemical reactions are not as random as they may appear.

This means that chimps would have a much better chance of typing something sensible on an advanced GPS screen than on an old typewriter, not because they are chimps but because of the design of the GPS. Let me use the analogy of a sophisticated GPS system in a car. When you hit the next character on the touch screen to enter the name of a city or a street, the screen shows or highlights only letters that qualify. Certain combinations of two letters are very common, but some are very rare or even impossible. Just as one can track frequencies of individual letters, one

[133] W. B. DeMore and C. W. Jacobsen, "Formation of Carbon Trioxide in the Photolysis of Ozone in Liquid Carbon Dioxide," *Journal of Physical Chemistry* 73, no. 9 (1969): 2935–2938.

can track frequencies of two-letter combinations and find that different languages have different patterns of preferred ones. For instance, in English, there is usually no *f* after a *b* or a *c*. In other words, there is much less randomness involved than explanation 1 assumes. This means that the chances of assembling more complex molecules from simpler molecules is dramatically greater than originally projected.

Similarly, the universe was designed in such a way that certain compounds are more likely to be "spontaneously" formed than others. There are indeed compounds that seem to be "favorites" in nature. Carbon (C) is the "favorite" building block of the living world on Planet Earth, because of its capability to form long chains of interconnecting C-C bonds, which allows carbon to form an almost infinite number of organic compounds. So, it should not surprise us that carbon is an essential part of proteins and DNA; such is the way our world was designed.

According to explanation 3, the origin of life is not so much a miracle of repeated divine interventions as a miracle of the proper cosmic design. The cosmic design of the universe has the potentiality for life. The Belgian biochemist, cell-biologist, and Nobel laureate Christian de Duve describes the origin of life as follows: "The pathway followed by the biogenic process up to the ancestral cell was almost entirely preordained by the intrinsic properties of the materials involved, given a certain kind of environment or succession of environmental conditions."[134] What this means is that life may not be such an unlikely process to start once we have the right conditions and ingredients after a long time of preparation after the Big Bang. In other words, the "secret" of life can be found in the cosmic design.

---

[134] Christian de Duve, *Blueprint for a Cell: The Nature and Origin of Life* (Burlington, NC: Neil Patterson Publishers, 1991), 214.

What Threatened Life?

Not only was life promoted by certain properties of its components and by certain conditions of Planet Earth; it was also threatened time after time. The best-known examples of these threats are mass extinctions and ice ages during the development of life.

Mass Extinctions

Mass extinctions, an unavoidable part of life's history, are periods when abnormally large numbers of species died out simultaneously or within a limited time. What could be characterized as destructive events, however, turned out to be constructive events that opened Planet Earth up for new forms of life.

Although the best-known mass extinction is the Cretaceous-Tertiary (or K-T) extinction event, which wiped out the dinosaurs, some other mass extinctions were even more devastating than the K-T. We know of at least five big mass extinctions that wiped out more than half of all species. As a result, more than 90 percent of all organisms that have ever lived on Earth are now extinct.

The first big mass extinction, and the third largest in Earth's history, happened some 450 million years ago. It's called the Ordovician-Silurian mass extinction. It had two peak dying times separated by hundreds of thousands of years. This was probably caused when the supercontinent at that time migrated into the South Pole region, which caused extensive glaciation and a dramatic drop in sea level as more water froze. Since most life was in the sea at that time, the event took its hardest toll on marine organisms such as brachiopods (similar in looks to clams), conodonts (resembling eels), and trilobites (similar in looks to horseshoe crabs). About 85 percent of all marine species disappeared.[135]

---

[135] Peter M. Sheehan, "The Late Ordovician Mass Extinction," *Annual Review of Earth and Planetary Sciences*, 29 (May 2001), 331–364.

In the Beginning

The second big mass extinction, the Late Devonian, happened about 375 million years ago. This drawn-out event eliminated about 70 percent of all marine species from Earth over a span of perhaps twenty million years. Life in the shallow seas was the most affected, and coral reefs took a hammering, not returning to their former glory until new types of coral evolved more than 100 million years later.

The third big mass extinction, the Permian-Triassic, happened some 250 million years ago. It has been nicknamed "the Great Dying," since the event was the deadliest of them all: more than 96 percent of all species perished.[136] All life on Earth today is said to have descended only from the 4 percent of species that survived. Many scientists believe that an asteroid or a comet triggered the massive die-off, but no crater has been found. Another strong contender is flood volcanism from a Siberian province in Russia. Impact-triggered volcanism is yet another possibility. It was during the Triassic recovery that conditions for life improved again. That's when dinosaurs and mammals first appeared. This happened when the supercontinent Pangea had split into two large continents.

The fourth big mass extinction, the Triassic-Jurassic, took place some 200 million years ago and included two or three phases of extinction. Massive floods of lava erupting from the central Atlantic magmatic area may explain this event. About 20 percent of all marine families became extinct, as well as most mammal-like creatures, many large amphibians, and all non-dinosaur archosaurs. An asteroid impact is another possible cause of the extinction, though here, too, a telltale crater has yet to be found. After this extinction, during the Jurassic recovery, large-bodied dinosaurs

---

[136] Sarda Sahney and Michael J. Benton, "Recovery from the Most Profound Mass Extinction of All Time," *Proceedings of the Royal Society B* 275 (April 2008), 759–765.

appeared on Earth. Their appearance was so rapid that scientists still have no explanation for this explosion.

The fifth big mass extinction, the Cretaceous-Tertiary, happened some 65 million years ago and is famed for the demise of the dinosaurs. Many other organisms also perished, including ammonites (which have shells with a spiral shape), many flowering plants, and the last of the pterosaurs (flying reptiles), which made room for mammals to diversify and evolve rapidly. An extraterrestrial impact is most closely linked to this event. A huge crater off Mexico's Yucatán Peninsula is dated to some 65 million years ago, coinciding with the extinction.[137] The impact induced multiple earthquakes of magnitude 11 or greater and unleashed tsunamis that crested above 100 meters. It unleashed more than 550 billion tons of sulfur into the atmosphere, which shut down photosynthesis for years.[138] But the recovery from this period replaced small-bodied mammals with a vast array of larger-bodied mammals.

These mass extinctions may seem rather haphazard, but in 1984 it was first noted that they had a nonrandom nature.[139] Subsequent research revealed that these events had an apparent 27-million-year periodicity over the last 500 million years.[140] This is rather consistent with the dates for Earth's impact craters

---

[137] Paul R. Renne et al., "Time Scales of Critical Events around the Cretaceous-Paleogene Boundary," *Science* 339 (February 2013): 684–687.

[138] Peter Schulte et al., "The Chicxulub Asteroid Impact and Mass Extinction at the Cretaceous-Paleogenene Boundary," *Science* 327 (March 2010): 1214–1218.

[139] D. M. Raub, and J. John Sepkoski Jr., "Periodicity of Extinctions in the Geologic Past," *Proceedings of the National Academy of Sciences* 81 (February 1984): 801–805.

[140] Adrian L. Melott and Richard K. Bambach, "Nemesis Reconsidered," *Monthly Notices of the Royal Astronomical Society Letters* 407 (September 2010): L99–L102.

larger than 35 kilometers in diameter, which show a 35-million-year periodicity.[141] A very likely explanation is that Earth resides in a planetary system with unique asteroid and comet belts. Earth has the best possible orbital path in the solar system to receive the kinds and frequency of asteroid and comet collisions that would "pace" Earth's mass extinction events.[142] Hugh Ross concludes from this that, had these events been totally random, the atmospheric heat-trapping capacity might have easily gotten out of sync with the sun's changing luminosity, enough to sterilize Earth permanently.[143]

For us nowadays who passionately focus on conservation of endangered species, it must be hard to deal with any kind of extinction, especially when it happens on a massive scale—when 50 to more than 90 percent of all species on Earth disappeared in a geological blink of the eye. There is another way of looking at them, however. Although these mass extinctions were deadly events, they opened up the planet for new life-forms to emerge. When older species faded away, new species had a chance to evolve and adapt to the ever-changing new environmental conditions.

So, mass extinctions could be seen as constructive rather than destructive. They were like stepping stones in the evolution of life, leading from unicellular organisms to multicellular organisms, to vertebrates, to mammals, and ultimately to human beings. Undoubtedly, they were drastic events in life history, but—and this is the most amazing part—never did Planet Earth become sterile

---

[141] Richard B. Stothers, "The Period Dichotomy in Terrestrial Impact Crater Ages," *Monthly Notices of the Royal Astronomical Society* 365 (January 2006): 178–180.

[142] Gennady G. Kochemasov, "On the Uniqueness of Earth as a Harbor of Steady Life: A Comparative Planetology Approach," *Astrobiology* 7 (June 2007): 518.

[143] Ross, *Improbable Planet*, 169–171.

during this period of at least five mass extinctions,[144] which ran from December 15 to 28 on the scale of one year.

Ice Ages

Ice ages are periods when the entire Earth experiences notably colder climatic conditions. During an ice age, the polar regions are cold; there are large differences in temperature from the equator to the pole; and large, continent-size glaciers can cover enormous regions of the earth. You probably know that water expands when it freezes—that's why icebergs float. This is a very fortunate law of nature. If ice sank, all the oceans would freeze solid and life would be impossible.

Planet Earth seems to have three main settings: "greenhouse," when tropical temperatures extend to the poles and there are no ice sheets at all; "icehouse," when there is some permanent ice, although its extent can vary greatly; and "snowball," in which the planet's entire surface is frozen over. This may put the current discussion about global warming in a wider perspective.

There have been five or six major ice ages over the past three billion years, resulting in the presence or expansion of continental and polar ice sheets and mountain glaciers. Do not think of an ice age as one long period of permanent ice. Within ice ages, there exist periods of more severe glacial conditions and more temperate ones. Within an ice age (or at least within the current one), more temperate and more severe periods alternate. The colder periods are called "glacial" periods, and intermittent warm periods are called "interglacial" periods.

These global cooling periods begin when a drop in temperature prevents snow from fully melting in some areas. The bottom

---

[144] Ross, *Improbable Planet*, 17.

layer turns to ice, which becomes a glacier as the weight of accumulated snow causes it to move slowly forward. A cyclical pattern emerges in which the snow and ice traps Earth's moisture, fueling the growth of these ice sheets as the sea levels simultaneously drop.

Earth is in the midst of an ice age, as the Antarctic and Greenland ice sheets remain intact despite moderate temperatures, but it is an interglacial period. The current ice age began 34 million years ago, with its latest phase in progress since 2.6 million years ago.

The oldest ice age we know about happened when Earth was just over 2 billion years old and home only to unicellular life-forms. Its early stages, from 2.4 to 2.3 billion years ago, seem to have been particularly severe, with the entire planet frozen over, causing the first "snowball Earth." This may have been triggered by a 250-million-year lull in volcanic activity, which would have meant that less carbon dioxide was being pumped into the atmosphere, resulting in a reduced greenhouse effect.

During the 200 million years of the second ice age (from 850 to 630 million years ago), Earth was plunged into some of the lowest temperatures it has ever experienced. This may have been caused by the emergence of more complex life-forms. One theory is that the glaciation was triggered by the evolution of large cells, and possibly also multicellular organisms, that sank to the seabed after dying. This would have sucked carbon dioxide out of the atmosphere, which weakened the greenhouse effect and thus lowered global temperatures. But speculation abounds.

The third ice age (from 360 to 260 million years ago) may have been the result of the expansion of land plants that followed the second ice age. As plants spread over the planet, they absorbed carbon dioxide from the atmosphere and released oxygen through the process of photosynthesis. As a result, carbon dioxide levels fell and the greenhouse effect weakened, triggering an ice age.

The fourth ice age (from 34 to 3 million years ago) began when the first small glaciers formed on the tops of Antarctica's mountains. And it was 20 million years later that worldwide temperatures dropped by 8 degrees Celsius. This temperature drop was triggered by the rise of the Himalayas as a result of plate tectonics. As they grew higher, they were exposed to increased weathering — the breaking down of rocks, soil, and minerals — which sucked carbon dioxide out of the atmosphere and reduced the greenhouse effect.

The fifth ice age started some 3 million years ago — and is still going on. So, its history is relatively recent, in geological terms, and can be studied in far more detail than the histories of the others. It's evident that the ice sheets have gone through multiple stages of growth and retreat over the course of this ice age. The main trigger for this glaciation was the continuing fall in the level of carbon dioxide in the atmosphere due to the impact of the Himalayas. The timing of the glacial and interglacial periods, however, was driven by periodic changes in Earth's orbit, which changed the amount of sunshine reaching various parts of the planet. During the first two-thirds of this ice age, the ice advanced and retreated roughly every 41,000 years — the same pace as the changes in the tilt of Earth's axis.[145] About a million years ago, the ice switched to a 100,000-year cycle for reasons that are not quite clear.

As the fifth glacial period drew to a close and temperatures began to rise, there were two final cold snaps. First, the chilly "Older Dryas" of 14,700 to 13,400 years ago transformed most of Europe from forest to tundra, similar to modern-day Siberia. After a brief

---

[145] J. D. Hays, John Imbrie, and N. J. Shackleton, "Variations in the Earth's Orbit: Pacemaker of the Ice Ages," *Science* 194, no. 4270 (December 10, 1976): 1121–1132.

respite, the "Younger Dryas," between 12,800 and 11,500 years ago, froze Europe solid within a matter of months. Some have blamed a cometary impact, but there are other explanations.

Scientists have tried hard to explain the onset of these ice ages. At least the following factors seem to be important:

1. The composition of the atmosphere, especially the concentrations of greenhouse gases, such as carbon dioxide and methane
2. The motion of tectonic plates, resulting in changes in the relative location and amount of continental and oceanic crust on the earth's surface
3. The impact of relatively large meteorites
4. The impact of volcanism
5. Changes in the earth's orbit around the sun

Let's look at each of them briefly.

The Composition of the Atmosphere

There is evidence that greenhouse gas levels fell at the start of ice ages and rose during the retreat of the ice sheets, but it is difficult to establish cause and effect. Changes in Earth's atmosphere, especially the concentrations of greenhouse gases, may alter the climate, while climate change itself can change the atmospheric composition — such as by changing the rate at which weathering removes carbon dioxide. It has been proposed that the Tibetan and Colorado Plateaus, for instance, were immense carbon dioxide "scrubbers" with a capacity to remove enough carbon dioxide from the global atmosphere to be a significant causal factor of the 40-million-year cooling trend.[146]

---

[146] W. F. Ruddiman and J. E. Kutzbach, "Plateau Uplift and Climate Change," *Scientific American* 264, no. 3 (1991): 66–74.

The Motion of Tectonic Plates

A significant trigger in initiating ice ages is the changing positions of Earth's ever-moving continents. These changes affect ocean and atmospheric circulation patterns. When plate tectonic movement causes continents to be arranged such that a warm water flow from the equator to the poles is blocked or reduced, ice sheets may arise and set another ice age in motion. Today's ice age, for instance, most likely began when the land bridge between North and South America (the Isthmus of Panama) formed and ended the exchange of tropical water between the Atlantic and Pacific Oceans, significantly altering ocean currents.

The Impact of Relatively Large Meteorites

When a huge meteor collided with Earth about 2.5 million years ago and fell into the southern Pacific Ocean, it not only could have generated a massive tsunami but also may have plunged the world into an ice age.[147] Unfortunately, since this was a deep-ocean collision, there's no obvious giant crater to investigate, as there would have been if it had hit a landmass. Recently, the discovery of a 3.3-million-year-old meteorite impact site in Argentina has revealed another potential trigger for an ice age. The meteorite was at least 1 kilometer wide.[148] By comparison, the meteorite that caused the global extinction of the dinosaurs was 10 kilometers wide.

---

[147] James Goff, Catherine Chagué Goff, Michael Archer, Dale Dominey Howes, Chris Turney, "The Eltanin Asteroid Impact: Possible South Pacific Palaeomegatsunami Footprint and Potential Implications for the Pliocene-Pleistocene Transition," *Journal of Quaternary Science* (September 2012).

[148] R. B. Firestone et al., "Evidence for an Extraterrestrial Impact 12,900 Years Ago That Contributed to the Megafaunal Extinctions and the Younger Dryas Cooling," *Proceedings of the National Academy of Sciences*, 104, no. 41 (October 2007): 16016–16021.

In the Beginning

The Impact of Volcanism

The shifting of the earth's plates creates large-scale changes to continental masses, which, in turn, triggers volcanic activity that releases carbon dioxide into the air. It has also been suggested that undersea volcanoes released methane and thus caused a large, rapid increase in the greenhouse effect.

Changes in the Earth's Orbit around the Sun

This is arguably the most important factor. The earth's orbit around the sun varies between nearly circular and mildly elliptical; this change goes through a cycle of around 100,000 years. When the orbit is more elongated, there is more variation in the distance between the earth and the sun, and in the amount of solar radiation, at different times of the year.

In addition, the rotational tilt of the earth (its obliquity) changes slightly. A greater tilt makes the seasons more extreme. The tilt of the earth's axis fluctuates by about two degrees in a 41,000-year cycle. Moreover, the earth's axis spirals in a cycle of 26,000 years, much like a spinning top. The spinning of the earth's axis and the elliptical rotation of the axis cause the day on which the earth is closest to the sun to migrate through the calendar year in a cycle of about 20,000 years: currently, it is at the beginning of January; in about 10,000 years, however, it will be at the beginning of July.[149]

Interestingly, predicted changes in the tilt of the earth's axis and the shape of the earth's orbit—which would alter the total amount of sunlight reaching Earth by up to 25 percent at middle

---

[149] E. T. H. Zurich, "Why an Ice Age Occurs Every 100,000 Years: Climate and Feedback Effects Explained," *Science Daily*, August 7, 2013.

latitudes—suggest that the next glacial period would begin at least 50,000 years from now.[150]

What we see here again is that although the occurrence of ice ages, like the occurrence of mass extinctions, may seem coincidental, or even chaotic, events, they always happen against a background of law and order. Somehow, they shaped Planet Earth the way we know it now: as a home for us.

The Dawn of Humanity

Apparently, Planet Earth was finally ready to be a home for us human beings. One significant outcome of the most recent ice age was the development of *Homo sapiens*. Humans were able to adapt to the harsh climate by developing such tools as the bone needle to sew warm clothing, and they used the land bridges to spread to new regions. What is beyond doubt is that *Homo sapiens* survived and turned to farming soon after the ice retreated, setting the stage for the rise of modern civilization. Then, by the start of the warmer era, humans were in position to take advantage of the more favorable conditions by developing agricultural and domestication techniques. Meanwhile, giants such as mastodons and mammoths, saber-toothed cats, giant ground sloths, and other megafauna that had reigned during the glacial period went extinct by its end.

What is it that makes human beings so different and so unique compared with other life-forms? There are at least several features that stand out. Human beings have faculties we find nowhere else in the living world. Let's briefly discuss them.[151]

---

[150] A. Berger and M. F. Loutre, "Climate. An Exceptionally Long Interglacial Ahead?" *Science* 297, no. 5585 (August 2002): 1287–1288.

[151] More on this in Gerard M. Verschuuren, *At the Dawn of Humanity—The First Humans* (Kettering, OH: Angelico Press, 2019).

In the Beginning

Language

Human beings are able to use *language*. Animals may be able to utter sounds, but that doesn't mean they use language. Animals may be able to learn commands, but that doesn't mean they understand language. Experts in the field of linguistics pointed out rather early that the simple "sentences" that apes such as Washoe, Koko, and Nim were taught to utter during experiments turn out to lack any kind of grammar—they are at best something like "Me Tarzan, you Jane."

In contrast, the language faculty in the human species is ubiquitous (apart from pathological exceptions). A newborn human instantly selects from the environment language-related data, whereas an ape, with the same auditory system, picks up only noise.

What makes language so different then? Many people think that using language, either in speaking or in writing, is merely a matter of putting words together in a certain order. Most words, however, are only "labels" for something else: *concepts*. Words are elements of language, but concepts are elements of thought.

Take the concept of "circle." After having seen several round objects, humans can abstract from this the concept of circle. This concept is abstract—in this case, even highly abstract. It is very unlikely that we ever encounter a perfect circle in this world, which means we do not literally or physically see a circle. Besides, the concept of circle does not include any specific size, whereas the circular objects around us do. True, we can visualize a circle without imagining any specific size, but concepts have a universality that images can never possess.

As the legendary philosopher Ludwig Wittgenstein explained, pointing at things is not sufficient to define words and their underlying concepts.[152] Of course, I can explain what the word *red* means

---

[152] Ludwig Wittgenstein, *Philosophical Investigations*, §§258–277.

by pointing at a red tulip. But that gesture is still very ambiguous. Perhaps someone might think *red* stands for a tulip, or for a flower, or whatever else might come to mind in connection with pointing at a red tulip. Besides, for many concepts, there is nothing to point at. To explain the concept of "tomorrow," for instance, there is nothing to point at (other than a calendar).

No wonder concepts play a central role in how we know the world. Thanks to concepts, we can see similarities that are not immediately visible and not directly tied to what we perceive. Everyone can see things falling, but to perceive "gravity," one needs the concept of gravity to see what no one had been able to see before Isaac Newton. The concept of gravity allows us to "see," for example, the similarity between the motion of the moon and the fall of an apple. Animals, on the other hand, can see things falling, but they don't see gravity.

In their thoughts, humans do not deal with things directly but after making a "detour" through concepts; they extrapolate from what is seen to what is unseen; they can assign various interpretations from different perspectives to the physical things they see around them. They move from the world of sensible singulars, physical things, to the world of immaterial universals, concepts, and symbols.

That is why the philosopher Ernst Cassirer suggested calling *Homo sapiens* a symbol-making animal (*animal symbolicum*). Humans can create mental concepts that transform things of the world into objects of knowledge. Thus, they can see with their "mental eyes" what no physical eyes could ever see before. In a sense, concepts help us to illuminate what was in darkness before. Science is a master at that.

It should not come as a surprise, then, that language is not primarily seen as a system of communication but rather as a system of *thought*—which can also, but only secondarily, be used for communication.[153] It was only at some later stage of evolution that the internal language of thought was connected to the sensorimotor system. That's when thought became connected to speech. Instead of starting as a communication system, language evolved primarily as a "tool for thought." Humans are preeminently thinking creatures, even in their language.

Rationality

Humans are rational beings, which means they can put concepts together in a rational way in statements, arguments, and the like. In fact, it is the faculty of rationality that gives us access to the world of truths and untruths—a world beyond our control. Rationality is our capacity for abstract thinking and having reasons for our thoughts, thus giving us access to the unseen world of thoughts, laws, and truths. Rationality allows us to gain knowledge about the world through the power of abstract concepts and mental reasoning, thus giving us an immaterial sense for what is true and what is false. Weighing evidence and coming to a conclusion are rational activities par excellence. Science could not exist without the faculty of rationality. Each time we look for any kind of explanation, we are in search of some form of reasoning.

Laws of nature are also the outcome of rationality. Unlike all material things surrounding us, scientific laws of nature do not have any of the features that apply to the material world. Yet when scientists or engineers violate these "immaterial" laws, they get themselves into real trouble. A bridge that has been designed according to the right laws can stand firm, whereas another bridge

---

[153] Most notably and convincingly argued in Robert Berwick and Noam Chomsky, *Why Only Us: Language and Evolution* (Cambridge, MA: MIT Press, 2016).

collapses because its engineers erred in their calculations; perhaps they had the wrong laws in mind, or at least the wrong thoughts. If those laws and thoughts were only creations of the human mind, then the construction of a bridge could never depend on the right laws and the right thoughts. It makes no sense to say that competent engineers have better mental habits than their inept colleagues—they have better knowledge.

Morality

Human beings have the faculty of *morality*. They have this exceptional characteristic of using moral standards to judge human conduct as being morally right or wrong. Having moral judgments seems universal in humanity. In that sense, *Homo sapiens* is also *Homo moralis*. Humans ought to do what they ought to do. They evaluate actions as good or evil. They "intuitively" know that they have certain moral *duties* toward other human beings and that other human beings owe them certain moral *rights*. This intuitive knowledge comes with their faculty of morality. (Let me stress, though, that the term *morality* in this book is not another word for social behavior. These two are very different notions. Whereas social behavior does have evolutionary roots in the animal world, morality does not.)

How could one possibly argue that there is morality in the animal world? Well, it is rather obvious that animals do not have a moral code that tells them what is right and what is wrong, or what they owe other animals and what other animals owe them. They have no moral code to control their drives, lusts, instincts, and emotions. They just follow whatever "pops up" in their brains—and no one has the right morally to blame them. For example, we will never arrange court sessions for grizzly bears that maul hikers, because we know that bears are not morally responsible for their actions. Since animals have no moral code, they have no duties, no responsibilities, and consequently no rights. If animals really had moral rights, other animals would have the moral duty to respect and honor those rights as well.

In contrast, the faculty of morality is universal among human beings. It does not come with a specific race, nation, party, or church—it is a common property that belongs to all of us who belong to the species *Homo sapiens*. It is undeniable that, in addition to a genetic code, human beings have a moral code.

Some people like to reduce a moral code to a genetic code. These people, however, should ask themselves why we need a moral code to do what we would do or not do "by nature" anyway.[154] A morality that is supposedly preprogrammed in our genes would make a moral code completely redundant. We would all act right by mere nature, so it would not even be possible to do something morally wrong. This makes us realize there must be a moral code beyond and above a genetic code. We do need a moral code because God, according to St. Augustine, "wrote on the tables of law what men did not read in their hearts."[155]

Self-Awareness

Human beings have the faculty of self-awareness. Self-awareness is the ability to recognize oneself as an individual separate from one's surroundings and from other individuals. Self-awareness implies that I know that I remain the same person, even though my cells are constantly being replaced. Losing or replacing larger parts of my body may happen without losing my self. I can get a new heart, liver, and lungs; I can also get knee, hip, and ankle replacements; I can receive prostheses for hands and feet, arms and legs, and so on. Yet I know I am still my old self. I also know, for instance, that

[154] See R.C. Lewontin, "The Fallacy of Biological Determinism," *Sciences* (1976): 6–10.
[155] Augustine, *Enarrationes in Psalmos*, 57, 1.

I started at one point in life as a fertilized egg cell, and that at some point, I will be dead. It is my self that connects all the stages of my life as one long continuum.

There is something very enigmatic about this self. Self-awareness is a faculty that operates under the power of the mind, with its faculties of language, rationality, and morality. It is very hard, arguably impossible, to describe this power in purely material terms coming from physics, biology, neurology, or genetics. This makes it very questionable whether our understanding of the world can be entirely done by the material mechanism of the brain.

Instead, self-awareness has something to do with the mind, for it is hard to locate the awareness of self in the brain. It is easy to see how the *mind* can study the brain, but it is hard to see how the *brain* could ever study itself. Indeed, the brain can be studied by scientists, but it is the minds of the scientists, not the brains themselves, that do the studying. The physical world can never be studied by something purely physical, any more than DNA could ever discover DNA, or neurons could ever discover neurons all by themselves.

Whereas the world of my body is public and accessible to others, the world of my self is private. It's through introspection (and retrospection) that I have access to my private world in a way no one else has. Even brain scans do not have access to my private world, as I do; all they can pick up is brain waves, but never my thoughts, for those fail to show up on pictures and scans.

Of course, there have been attempts to link the faculty of self-awareness to the rest of the animal world, as an outcome of evolution. One of the tools that claims to prove this is the so-called mirror test. When first exposed to a mirror, most apes react to their reflections as they would to another member of their species; but eventually they may learn to recognize their mirror images as reflections of their own bodies. So it must be admitted that some animals can (learn to) recognize their own bodies.

In the Beginning

In animals, however, awareness of their bodies is not necessarily self-awareness. The fact that birds, for instance, can clean their feathers doesn't mean that they have the capacity of self-awareness. Besides, animals that regularly drink from water surfaces should be used to seeing a reflection of their bodies in the water anyway. But do they have a sense of self, a sense of past and future, knowing that they exist in a particular time and place? It is very doubtful they do, for what mirrors reflect are bodies, not selves. Animals don't go to the water just to look at themselves in the "mirror," whereas many humans tend to look in mirrors many times a day.

Another important feature of self-awareness is that it also comes with *death* awareness. This may open a window to the question of when and where the human self-emerged. If there is any indication of death awareness, we may assume there is also self-awareness. Scientists have no way of spotting death awareness, other than going by burial practices, which are an indication of death awareness—and therefore of self-awareness. Scientists consider evidence of burial rituals and the presence of grave goods an important indicator of being human, because these may signify self-awareness as well as a concern for the dead that transcends daily life.

Burial of the dead with material goods indicates a belief in an afterlife, for the goods are seen as useful to the deceased in their future lives, where their selves may go on. The earliest, virtually undisputed human burial dates back some eighty thousand years, in the Skhul cave at Qafzeh, Israel.[156] It is very likely that we are dealing here with human beings who had a sense of self. The legendary biologist Theodosius Dobzhansky could not have said it

---

[156] Erik Trinkaus, "Femoral Neck-Shaft Angles of the Qafzeh-Skhul Early Modern Humans, and Activity Levels among Immature Near Eastern Middle Paleolithic Hominids," *Journal of Human Evolution* 25, no. 5 (November 1993).

better: "A being who knows that he will die, arose from ancestors who did not know."[157]

Once the first humans arrived, we find evidence not only of elaborate burials for the first time in prehistory but also of aesthetic expressions. Findings in prehistoric graves suggest that decoration and art were an integral part of the lives and societies of the people who made them. That means that these humans were masters not only of self-awareness but also of self-expression. Of all animals in prehistory, only humans left stones behind with inscriptions; only humans created paintings in caves, and so forth. As a matter of fact, we have found astonishing art in prehistoric caves. Some of the first Cro-Magnon sites, dating from well over thirty thousand years ago, have even yielded evidence of music: multi-holed bone flutes capable of producing a remarkable complexity of sound.

Self-Transcendence

Human beings have the faculty of self-transcendence. When I say, "I am only human," I am not comparing myself with something "below" me (such as a cat, a dog, or a chimp), but I am comparing myself with Someone who is "above" me and who transcends me. When I call myself "only human," I am comparing myself with a Person who does not have the limitations I experience in myself. In some mysterious way, I am reaching out into the realm of the absolute, far beyond myself. In doing so, the finite catches a glimpse of the Infinite. That's a major characteristic of all religions: assuming some form of self-transcendence and knowing that there is Someone who transcends my self—who is larger than I am, who created me, and who sustains me. Earlier we spoke of a First Cause. Belief in the Transcendent is at the heart of all religions; it unites

---

[157] Theodosius Dobzhansky, "Changing Man," *Science* 155 (January 1967): 68.

the orthodox forms of Judaism, Christianity, Islam, Buddhism, and even Hinduism.

St. Thomas Aquinas is famous for putting this thought of an absolute, infinite Being—the very core of religion—into philosophical terms. His reasoning starts as follows: we *receive* our existence; we are *contingent* and could easily not have been; we don't have to exist, but because we do exist, we can ask for the cause of our existence—God. God alone is the act of existing itself. If God ceased to hold each one of us in existence, we would simply disappear—we would be annihilated, returning to what we came from, nothing (*nihil*), at the beginning of creation.

The five faculties mentioned above set human beings apart from the rest of the animal world. As a matter of fact, it is very hard, even impossible, to explain these faculties in purely scientific terms. Invoking neurons, genes, and the like to explain them cannot do the job, for the simple reason that, for science to do so, it depends on concepts, which are immaterial entities. Thoughts are not something the brain secretes. Reducing concepts to a "creation of neurons in the brain" obscures the fact that "neuron" itself is an abstract, immaterial concept. Such a claim starts a vicious circle, which is a sign there is something seriously wrong.

Stephen Barr shows us the vicious circle: "The very theory which says that theories [or concepts] are neurons firing is itself naught but neurons firing."[158] Something similar can be said about genes. If concepts are merely a product of genes, we are reasoning in circles again, because the concept of gene is an immaterial, abstract concept in itself. Consequently, we will never be able to lift concepts like these off the ground—they depend on the very concepts we are trying to explain. We are, in fact, trying to prove something we had started with.

---

[158] Barr, *Modern Physics and Ancient Faith*, 196.

If these distinctive human features did not come from our prehuman ancestors nor from their prehuman genes or prehuman brains, then the question arises: Where could these distinctive human features have come from? The Christian answer is rather brief: from their immortal souls.

And their immortal souls come directly from God.

The human mind is the intellectual part of the human soul. If the mind is identical to the brain, and if mental issues are nothing but neural issues, then the discussion is closed, and everything that happened to the first humans can be and should be fully explained in terms of their anatomy, their physiology, and their genetics. But if the mind is not the same as the brain, then the discussion is far from over. It must be more of an enigma than science suggests.

The emergence of humanity with its faculties of language, rationality, morality, self-awareness, and self-transcendence—most likely some eighty thousand years ago[159]—was a climax in the evolution of life on Planet Earth. It happened at the last minute—at 11:59 p.m. on the scale of a day. We may certainly speak of a climax, for never since has a new species arisen with the same faculties as the species *Homo sapiens*, let alone more faculties. How did the universe get to this culmination?

We don't fully know, certainly not in purely scientific terms. Perhaps the best we can say is that the road to humanity is a process that meanders like a river. On the one hand, it follows a path that seems coincidental and random. On the other hand, in spite of its winding flow, it also moves in a specific direction, steered by a path of least resistance. A river follows a path of least resistance, according to the topographic design of the landscape. In a similar way,

---

[159] Christopher Henshilwood, et al., "Emergence of Modern Human Behavior: Middle Stone Age Engravings from South Africa," *Science* 295 (2002): 1278–1280.

the stream of evolution follows a path regulated by what we called earlier the cosmic design of our universe. In other words, evolution follows the path of least resistance in the landscape of the cosmic design. It does not just flow in a purely random manner. The cosmic design creates the "bed" in which the stream of evolution meanders. Somehow, the flow of evolution found its destination in humanity.

8

Were We Meant to Be Here?

Human beings may be very special and unique creatures, but that doesn't necessarily mean Planet Earth was made with them in mind. Or does it?

Science May Not Have an Answer

Most scientists believe that we are *not* meant to be here. The main reason for thinking so is that science has the last word, in their opinion, for only science has the power to verify or falsify whatever we claim. In their view, only science can prove that we were meant to be here or not meant to be here. But that's exactly the point of contention. Does science really have the tools to decide whether we were meant to be here? If it does, we have to ask science that question and wait for its answer. But if it doesn't, we cannot expect any definitive answer from science for or against the question of whether we are meant to be here.

Most scientists have already taken a position in this dispute. They probably consider much of what we discussed in this book to be unscientific, and therefore useless. They would tell us that all the claims made here should be rejected. The question: What makes them think so? The most common reason given for this rejection is that science should always have the last word. Why?

It is their conviction that the scientific method is not only the best method there is, but also the only method we have to understand the universe. This makes them believe that all our questions have scientific answers phrased in terms of particles, quantities, and equations. They glorify what science studies—matter, that is, which can be dissected, counted, quantified, and measured. In other words, they have a dogmatic, unshakable belief in the omnicompetence of science.

This conviction has been expressed in many ways. Someone like Richard Dawkins probably speaks on behalf of all of them when he says, "Religions still make claims about the world that on analysis turn out to be scientific claims."[160] This then gives Dawkins reason to say, "Gaps shrink as science advances, and God is threatened with eventually having nothing to do and nowhere to hide."[161] He once said to someone in his audience who asked about the existence of God, "Oh, all sorts of funny things happen in people's heads. But you can't measure them, so they do not mean anything."[162]

Another outspoken atheist is the chemist and Nobel laureate Peter Atkins, who thinks that scientists are privileged "to see further into truth than any of their contemporaries.... There is no reason to expect that science cannot deal with any aspect of existence."[163] Notice the word *any*. Scientists have made many

[160] Richard Dawkins, *A Devil's Chaplain* (Phoenix: Orion, 2003), 150.
[161] Richard Dawkins, *The God Delusion* (Boston: Houghton Mifflin, 2006), 147.
[162] Quoted by Francis Spufford, "Spiritual Literature for Atheists," *First Things* (November 2015).
[163] Peter Atkins, in *Nature's Imagination: The Frontiers of Scientific Vision*, ed. J. Cornwell (Oxford: Oxford University Press, 1995), 123.

other such dogmatic statements. Here is one more: Francis Crick, the codiscoverer of DNA, made his opinion very clear: "Almost all aspects of life are engineered at the molecular level, and without understanding molecules we can only have a very sketchy understanding of life itself." At least he had the intellectual honesty to acknowledge that it's "almost all aspects of life." I thank him for the word *almost*.

We should seriously question whether claims like these can be true. If they are true, then many claims made in this book are nothing more than illusions or delusions. So, we need to ask: Does science really have the last word? Or is that claim a form of megalomania that proclaims science to be a know-all and cure-all? Is it merely a claim coming from the dictatorship of what's called "scientism"?

As a matter of fact, there are many reasons why *scientism* is a form of misplaced megalomania. Let's see why.

There is nothing wrong with defending science, but there is much wrong with defending scientism—the notion that science is all there is. Science may be everywhere, but science is certainly not all there is. People who preach scientism are crusaders for a dubious case. How can they defend scientism, since scientism itself does not follow its own rule? How can science ever prove by itself that science is the only way of finding truth? There is no experiment that can do the trick. Science cannot pull itself up by its own bootstraps any more than an electric generator is able to run on its own power.

As a matter of fact, one cannot talk *about* science without stepping *outside* science. Scientism must step outside science to show that there is nothing outside science and that there is no other point of view—which does not seem to be a very scientific move.

Just think of the statement "No statements are true unless they can be proven scientifically," which in itself cannot be proven true scientifically. It is not a scientific discovery but at best a philosophical or metaphysical viewpoint—and a poor one at that. It declares everything outside science to be a despicable form of philosophy, in defiance of the fact that all those who reject philosophy are, in fact, committing their own version of philosophy. Scientism rejects religious faith and replaces it with its own "faith." This makes scientism a totalitarian ideology, for it allows no room for anything but itself.

But there is more that is inherently wrong with scientism. To claim that science has the final answers to our questions is factually mistaken. Science never has *final* answers. Science is a work in progress. What we call "proven" scientific knowledge is proven only until a new set of empirical data disproves it. In science, whatever is considered true today may be found false tomorrow. Francis Crick couldn't have said it more forcefully: "A theory that fits all the facts is bound to be wrong, as some of the facts will be wrong."[164] He could have expressed this more accurately, though, by stating that facts cannot be wrong, but they may turn out not to be facts. For a long time, it was a "known fact" that the earth is flat, but it has turned out *not* to be a fact.

As a matter of fact, there are many scientific theories that were believed to be true but then turned out to be false. Here is just a small selection of them:

1. The expanding-earth hypothesis stated that phenomena such as continental drift could be explained by the fact that the planet was gradually growing larger.[165] Now we know better.

---

[164] Francis Crick, *What Mad Pursuit: A Personal View of Scientific Discovery* (New York: Basics Books, 1990), 60.

[165] M. W. McElhinney, S. R. Taylor, and D. J. Stevenson, "Limits to the Expansion of Earth, Moon, Mars, and Mercury and to Changes in the Gravitational Constant," *Nature* 271 (1978): 316–321.

2. Before scientists embraced the notion that the universe was created as the result of the Big Bang, it was commonly believed that the size of the universe was an unchanging constant. The static universe is also known as "Einstein's Universe," for Einstein argued in favor of it and even calculated it into his theory of general relativity.[166]

3. For about thirty years, the number of human chromosomes was thought to be forty-eight, until the geneticist Joe Tjio found it to be forty-six in 1955.[167]

4. For a long time, stomach ulcers were believed to be caused by stress, until in 1982 Robin Warren proved them to be caused by a bacterium, *Helicobacter pylori*.[168]

5. For decades, it was a dogma in cell biology that the flow of genetic information is unidirectional—from DNA to RNA to protein—until it was discovered that RNA could convert back into DNA.[169]

6. Until the mid-twentieth century, most paleoanthropologists preferred Asia over Africa as the continent where the first hominids evolved. The recent African origin of modern humans, however, is the currently preferred theory (until further notice).[170]

---

[166] Cormac O'Raifeartaigh, Michael O'Keeffe, Werner Nahm, Simon Mitton, "Einstein's 1917 Static Model of the Universe: A Centennial Review," *European Physical Journal* 42, no. 3 (2017): 431–474.

[167] J. H. Tjio and A. Levan, "The Chromosome Number in Man," *Hereditas*, 42 (1956): 1.

[168] J. R. Warren, "Unidentified Curved Bacteria on Gastric Epithelium in Active Chronic Gastritis," *Lancet* 321, no. 8336 (1983): 1273–1275.

[169] David Baltimore, "RNA-Dependent DNA Polymerase in Virions of RNA Tumor Viruses," *Nature* 226, no. 5252 (1970): 1209–1211.

[170] Chris Stringer, "Human Evolution: Out of Ethiopia." *Nature* 423, no. 6941 (June 2003): 692–695.

Nevertheless, it remains a timeless temptation to claim the possibility of certainty and finality in science. But is this dream realistic? Most people accept that certainty and finality may "not yet" be possible in "softer" sciences such as biology and psychology, but as a matter of fact, it may not be possible even in the "hard" science of physics. The Dutch physicist Pieter Zeeman, later to become a Nobel laureate, was fond of telling how in 1883, when he had to choose what to study, people had strongly dissuaded him from studying physics. "That subject's finished," he was told, "there's no more to discover."[171]

It is even more ironic that this also happened to Max Planck, since it was he who, in 1900, laid the foundations for one of the greatest leaps in physics, the quantum revolution.[172] Fortunately, some years ago, Stephen Hawking ended his inaugural lecture on a more realistic note: "Though we know what end we are looking for, that end does not seem to be in sight for quite a long while to come."[173] Yet there is this timeless temptation to claim that the unknown has been reduced to almost nothing. The magnitude of the unknown is, however, well ... unknown! Science is by its nature a work in progress. It does not seem to have *final* answers. Inevitably, this observation puts the scientific data given in this book in a different perspective, too. Sorry about that!

---

[171] Mentioned in A. Van Den Beukel, *The Physicists and God* (North Andover, MA: Genesis Publishing, 1996), 37.

[172] From a 1924 lecture by Max Planck, *Scientific American*, Feb 1996, 10. Also: Alan P. Lightman, The *Discoveries: Great Breakthroughs in Twentieth-century Science, Including the Original Papers* (Toronto: Alfred A. Knopf, 2005), 8.

[173] Stephen Hawking, *Is the End in Sight for Theoretical Physics? An Inaugural Lecture* (Cambridge: Cambridge University Press, 1980), 22.

Another reason to question scientism is the fact that its defenders proclaim that science has *all* the answers to *all* our questions. This belief should at least be seriously probed.

The late University of California at Berkeley philosopher of science Paul Feyerabend, for instance, comes to the opposite conclusion when he says, "Science should be taught as one view among many and not as the one and only road to truth and reality."[174] Even the "positivistic" philosopher Gilbert Ryle expressed a similar view: "The nuclear physicist, the theologian, the historian, the lyric poet and the man in the street produce very different, yet compatible and even complementary pictures of one and the same 'world.'"[175] Science provides only one of these views.

But the list of problems that scientism has to face does not end here. As a matter of fact, a method as successful as the one that science provides does not disqualify any other methods. A blood test, for instance, is an excellent method to assess a person's health, but there are many other reliable methods, such as X-rays and MRIs, depending on what we are trying to assess. But a blood test on its own cannot be used to prove that a blood test is the best and only method there is.

Yet that is what scientism tries to do; it steps outside science and then claims, in an unscientific way, that science has the only legitimate method of knowing and offers us the only reliable view on the world. First, scientism declares a particular method, the scientific method (whatever that is), as far superior, and then claims that this disqualifies any other methods.

---

[174] Paul Feyerabend, *Against Method: Outline of an Anarchistic Theory of Knowledge* (New York: Verso Books, 1975), viii.

[175] Gilbert Ryle, *Dilemmas* (Cambridge, UK: Cambridge University Press, 1960), 68–69.

But there is much more to say against scientism. Scientific knowledge does not even qualify as a superior form of knowledge. It may be more easily testable (really?) than other kinds, but it is also very restricted and therefore requires additional forms of knowledge. Mathematical knowledge, for instance, is the most secure form of knowledge, but it is not about anything material. Other kinds of knowledge may arguably be more significant, but they are less secure. Einstein said it right: "As far as the laws of mathematics refer to reality, they are not certain; and as far as they are certain, they do not refer to reality."[176] So there must be many other forms of knowledge. Consider the analogy used by the philosopher Edward Feser: a metal detector is a perfect tool to locate metals, but that does not mean there is nothing more to this world than what metal detectors can detect.[177]

Those who protest that this analogy is no good, on the grounds that metal detectors detect only part of reality wheras physics detects the whole of it, are simply begging the question again, for whether physics really does describe the whole of reality is precisely what is at issue. An instrument can detect only what it is designed to detect. And that is exactly where scientism goes wrong: instead of letting reality determine which techniques are appropriate for which parts of reality, scientism lets its favorite technique dictate what is considered "real" in life—in denial of the fact that science has purchased success at the cost of limiting its ambition.

To characterize this restricted attitude of scientism, an image used by the late psychologist Abraham Maslow might be helpful: if you have only a hammer, every problem begins to look like a

---

[176] Address to Prussian Academy of Sciences, 1921.
[177] Edward Feser, *Scholastic Metaphysics: A Contemporary Introduction* (Heusenstamm, Germany: Editiones Scholasticae, 2014).

nail.[178] So, instead of idolizing our scientific hammer, we should acknowledge that not everything is a nail. Even if we were to agree that the scientific method gives us better testable results than other sources of knowledge, this would not entitle us to claim that only the scientific method gives us genuine knowledge of reality.

Admittedly, it is true that if science does not go to its limits, it's a failure, but it is equally true that, as soon as science oversteps its limits, it becomes arrogant—a know-it-all, a form of gross megalomania. No wonder this has led some to criticize scientism as a form of circular reasoning. The late philosopher Ralph Barton Perry expressed this as follows: "A certain type of method is accredited by its applicability to a certain type of fact; and this type of fact, in turn, is accredited by its lending itself to a certain type of method."[179] That's how we keep circling around.

As if that's not reason enough for rejecting scientism, we need to mention another problem scientism faces. No science, not even physics, is able to declare itself a superior form of knowledge. Some scientists have argued, for example, that physics always has the last word in observation, for the observers themselves are physical. But why not say, then, that psychology always has the last word, because these observers are interesting psychological objects as well? Neither statement makes sense; observers are neither physical nor psychological, but they can indeed be looked at and studied from a physical, biological, psychological, or statistical viewpoint—which is an entirely different matter.

Often scientism results from hyperspecialized training coupled with a lack of exposure to other disciplines and methods. This

---

[178] Abraham Harold Maslow, *The Psychology of Science* (New York: HarperCollins, 1966), 15.
[179] Ralph Barton Perry, *Present Philosophical Tendencies* (New York: Longmans, Green, 1912), 75.

may keep one from realizing that the findings of science are always partial and fragmentary. There is no science of "all there is." There may someday be a "grand unified theory" in physics—a theory that unifies the three nongravitational forces—but that is not the same as a "grand unified theory of everything." A theory of *everything* would also have to explain why some people believe in that theory and some do not.

A more profound argument against scientism is based on the fact that science is about material things, yet there is so much more in life than material things. Even science itself requires immaterial things, such as logic and mathematics. There is good and bad logic, good and bad math. But as G. K. Chesterton liked to ask his readers, "Why should not good logic be as misleading as bad logic? They are both movements in the brain of a bewildered ape?"[180] Yet, we know good logic or math is not as misleading as bad logic or math, although logic and mathematics are not physical, and therefore not testable by the natural sciences—and yet they cannot be ignored or denied by science. In fact, science heavily relies on logic and mathematics to interpret the data that scientific observation and experimentation provide.

In addition to immaterial things such as logic and mathematics, science even depends on immaterial assumptions that find their foundation in religion—assumptions such as order and intelligibility. Science cannot be done without a series of assumptions; they are "hidden," as most of an iceberg is hidden under water. The biologist Ernst Mayr speaks of "silent assumptions that are taken so completely for granted that they are never mentioned."[181]

---

[180] G. K. Chesterton, *Orthodoxy* (Chicago: Moody Publishers, 2009), 33.

[181] Ernst Mayr, *The Growth of Biological Thought* (Cambridge, MA: Harvard University Press, 1982), 835.

In setting the record straight against scientism, one could also bring in an argument of a historical nature. The first legendary pioneers of science in England were very much aware of the fact that there is more to life than science. When the Royal Society of London was founded in 1660, in its charter, King Charles II assigned to the fellows of the Society the privilege of enjoying intelligence and knowledge, but with the following important stipulation: "provided in matters of things philosophical, mathematical, and mechanical."[182] (*Philosophical* meant "scientific" back then.) Its members had their area of investigation explicitly demarcated, and they realized very clearly that they were going to leave many other domains untouched. That's how the domains of knowledge were separated; it was this "partition" that led to a division of labor between the sciences and other fields of human interest.

By accepting this separation, science bought its own territory, but necessarily at the expense of all-inclusiveness—the rest of the "estate" was reserved for others to manage. On the one hand, it gave to scientists all that could "methodically" be solved by dissecting, counting, and measuring. On the other hand, these scientists agreed to keep their hands off all other domains—education, legislation, justice, ethics, and certainly religion—because those would require a different expertise.

There are probably more reasons to reject scientism. But the ones mentioned above are arguably powerful enough to disqualify

---

[182] The Royal Society originated on November 28, 1660, when twelve men met to set up "a Colledge for the promoting of Physico-Mathematicall Experimentall Learning." Robert Hooke's draft of its statutes reads literally: "The Business and Design of the Royal Society is: To improve the knowledge of natural things, and all useful Arts, Manufactures, Mechanik practices, Engyries and Inventions by Experiments— (not meddling with Divinity, Metaphysics, Moralls, Politicks, Grammar, Rhetorik, or Logic)."

scientism, so it can no longer pose a serious threat to the claims made in this book.

Yet, we have to keep in mind that scientism is still very much alive, albeit mostly hidden underground, in spite of all the above objections. The late Dutch physicist Hendrik Casimir—the Casimir effect of quantum-mechanical attraction was named after him—said, "We have made science our God."[183] Indeed, science has become a semi-religion, in which the scientists are the new "priests." Science is supposed to explain *everything*, and in a much better way than God once did in their view. It was in this frame of mind that Stephen Hawking exclaimed, "Our goal is a complete understanding of the events around us and of our own existence."[184] Scientism likes to broadcast to everyone around, "It's all about science."

Well, science may be everywhere, but science is certainly not all there is—claiming differently would make for a shaky ideology. So, let's take courage and embark on an exciting journey through the rest of this book—a journey behind the scenes of the scientific data given in the first half of this book. If there is indeed more to life than science can give us, then let's go for that "more"—for what's beyond the scientific data.

The Cosmic Design

How did God make Planet Earth our home? Not by gambling with large numbers, not by making a series of divine interventions, but by creating a universe outfitted with what we called earlier the cosmic design. This cosmic design represents the order and structure of our universe, based on physical constants and laws

---

[183] Quoted in Van Den Beukel, *The Physicists and God*, 30.
[184] Hawking, *A Brief History of Time*, 186.

of nature that steer processes in the universe. And as such, it is the foundation our universe was built on — the matrix of all that exists. Even all seemingly random events run through the "bed" of this cosmic design.

The cosmic design had to be designed by an intelligent Creator. Therefore, it is an intelligent design, which should not be confused with the "intelligent design" of the ID theory, which we discussed earlier. The cosmic design determines what is possible and what is not possible in the universe. Planets could not exist without the law of gravity. Fish could not swim if they didn't follow hydrodynamic laws. Birds could not fly if they didn't follow aerodynamic laws. This even holds for the artificial constructions humans make: computers, for instance, could not work if their engineers didn't follow the right laws and calculations.

The cosmic design comes with all that is best for humanity, for the earth, and for the universe, and, as such, it is part of divine providence, which has us human beings in mind. Science can help us to reveal this cosmic order and investigate its details. Especially in physics, empirical laws were found to follow from deeper and deeper laws and principles. Chemical bonds, for instance, follow from laws of atomic physics, and these, in turn, flow from the laws of quantum electrodynamics. Deeper and deeper levels of laws have been uncovered. Science tries to keep doing this until it reaches, hopefully, some ultimate law of nature. As Stephen Barr puts it, "The deeper one goes the more orderly nature looks, the more subtle and intricate its designs."[185]

Even if physics could ever explain with a grand unified theory why the universe is the way it is, however, that's where the search of science would come to an end. There would be nothing beyond that point for science to pursue. Yet there would be a pivotal question

---

[185] Barr, *Modern Physics and Ancient Faith*, 81.

left as to where such an overarching law of the cosmic design came from. The fundamental question remains: Why does the universe have the nature it does, and why does it exist at all? There is no law of logic, not even mathematics, that says that the laws of this universe have to be what they turned out to be. Even Isaac Newton realized that God could have made the universe differently if He so wished. That's why, according to Newton, particles have certain forces and not others, or planetary orbits have certain parameters and not others.[186]

So here is the question again: Why is the cosmic design of the universe this way? Its explanation is not to be found in itself, because the universe is a contingent entity that could easily not have been—which calls for an ultimate explanation and ground beyond itself: an orderly Creator, a rational Lawgiver, and an intelligent Designer.

Only the existence of an orderly Creator explains why there is order in this world. Only the existence of a rational Lawgiver explains why there are laws of nature. Only the existence of an intelligent Designer explains why there is design in nature and why nature is intelligible. How could there be human minds, if the universe itself were mindless? The answer is that the laws of nature and nature's physical constants are part of a "preloaded" cosmic design, implemented by God the Creator.

At the Mercy of Natural Selection?

How can we apply all of this to the course of evolution? What does evolution have to do with the cosmic design? Interestingly, even Charles Darwin, one of the fathers of evolutionary theory, had to admit in a letter to Asa Gray: "I am inclined to look at

[186] Isaac Newton, *Opticks*, query 31.

everything as resulting from *designed* laws."[187] What laws are we talking about here?

Probably the most important law of evolution is the law of natural selection. Whether that law explains everything in evolution is not the issue here; most biologists agree that it's only part of the explanation. It is at least a law that plays an important role in the process of evolution, however, albeit in addition to other factors. It was Darwin's conviction, as he used to say, that evolution follows laws in the same way as planets and comets follow laws in the physical sciences. In his own words, "astronomers do not state that God directs the course of each comet and planet."[188] He was right: comets and planets just follow laws.

Why wouldn't the same hold for the process of evolution? If losing wing functionality (such as in penguins) is a regular evolutionary process, why wouldn't developing wing functionality be one, too? And let's not forget that birds must obey the laws of aerodynamics, as fish must follow the laws of hydrodynamics.

It's the law of natural selection that promotes "good" biological designs over "bad" designs, which increases species' survival rates and thus their frequency in later generations. What makes biological designs "good" or "bad"? There is ultimately only one explanation: it depends on what the cosmic design requires. The more an organism is adapted to its environment—making for a better biological design to solve the problems posed by the environment—the more likely this organism is to contribute to the composition of the next generations. It is basically a simple, rather straightforward law. Put in a catch phrase, it holds that "success breeds success."

---

[187] *The Correspondence of Charles Darwin* (Cambridge: Cambridge University Press, 1993), 224, italics added.
[188] In correspondence with geologist Charles Lyell in 1861.

In the Beginning

Unlike mutations—which randomly create the diversity that natural selection selects from—natural selection is a highly selective process. As a matter of fact, the term itself leads us almost inevitably to the obvious question: "Selection by whom or by what?" Even Darwin always felt uneasy about the term *natural selection.* The implication of a "selecting agent" looms large.

Darwin certainly tried to avoid this implication by saying that he had as much right to use metaphorical language as physicists do. "Who objects to an author speaking of the attraction of gravity as ruling the movements of the planets?" he asked. "Everyone knows what is meant and is implied by such metaphorical expressions."[189] Yet, in many passages, he refers to "nature," some kind of feminine deity, as the agent of selection. Eventually, his colleague Alfred Wallace convinced him to replace the term *natural selection* with Herbert Spencer's notion of survival of the fittest in the fifth edition of his famous book, *The Origin of Species.*

Nevertheless, the question of "selection by whom or by what" remains legitimate. After what we have seen so far, we probably have a better idea what the selecting factor or agent is. The answer could go like this. The universe has an overall set of restraints harnessing individual biological designs and making them "fit" or "successful" to a certain degree. Ultimately, it's the cosmic design. It determines, for instance, which bridges are successful and which biological designs are successful. As Stephen Barr puts it, "When order comes out it is only because order was put in; so we still have order to explain."[190] Without the cosmic design of the universe acting in the background, biological as well as technological designs could not work at all. A heart could not pump blood if it did not follow hydrodynamic laws; a bird's wing would not let the bird fly

---

[189] Darwin, *The Origin of Species*, 5th ed., 4, 93.
[190] Barr, *Modern Physics and Ancient Faith*, 81.

if it did not follow aerodynamic laws. It is the rules and laws of the cosmic design that restrict the range of possible end results.

Certain biological features of organisms are "successful" and "effective" in reaching their "goal" because they have a biological design that enables such a successful goal. This used to be called *teleology*—but that word is out of favor now with many biologists. Yet, if those biological features were not properly designed—not teleological, that is—they simply would not work. Leon Kass, a University of Chicago professor and physician, could not have worded it better: Organisms "are not teleological because they have survived; on the contrary, they have survived (in part) because they are teleological."[191] It is the cosmic design that regulates which biological designs "work," and thus are fit and successful in evolution. In other words, natural selection does not create the fit—it only selects what fits. It does not explain a fit but uses a fit in order to select. The fit is based on the cosmic design.

Biological designs presuppose an overarching, comprehensive, grand design of the universe, for the simple reason that natural selection must *assume* design but can neither *explain* nor *create* it, as we saw. The cosmic design is an intelligent design inherent in all of creation, rooted in autonomous secondary causes, physical constants, and laws of nature, which harness and steer everything in the universe—complex and not so complex. It is only because of the cosmic design of a Divine Designer that natural selection can do the work it did and is still doing.

In other words, the "fittest" are not defined by their survival—that would make for a tautology—instead they are defined by their design. Consequently, biological fitness is not an outcome of natural

---

[191] Leon R. Kass, M.D., "Teleology, Darwinism and the Place of Man: Beyond Chance and Necessity?" in *Toward a More Natural Science* (New York: Free Press, 1988), chap. 10.

selection but a condition for natural selection. Natural selection does not create teleology, but its working is based on teleology and presupposes teleology, for it cannot do its work without teleology.

Ironically, the atheist in biologist Richard Dawkins speaks of "a universe without design" in one of his books.[192] From this, he concludes that there is no longer any need in science for a cosmic Designer. At the same time, he keeps talking about successful biological designs that have made it in evolution. So, he is actually still talking about successful designs—but in his view, they are *biological* designs without any *cosmic design*. But the question remains: How can they be "successful" or "fit" if there is nothing that determines what they are "fit" for and what their "success" is?

To put it differently, the cosmic design acts as a filter in evolution. Although the atheist in the biologist Dawkins rejects the design concept, curiously enough, he says, "Each generation is a filter, a sieve; good genes tend to fall through the sieve into the next generation; bad genes tend to end up in bodies that die young or without reproducing."[193] What he seems to forget is that the terms *good* and *bad* already are design terms with a teleological connotation. And so are terms such as *successful* and *fit*; they are, in essence, teleological concepts connected to another teleological concept, "design." So, the real filter in evolution is actually the cosmic design of this universe—not natural selection. The cosmic design does not filter just randomly; it filters according to design criteria.

The philosopher Michael Augros makes a strong case for a designed universe. "If the first cause is intelligent," he reasons, "then

---

[192] Richard Dawkins, *The Blind Watchmaker: Why the Evidence of Evolution Reveals a Universe without Design* (New York: W. W. Norton, 2015).

[193] Richard Dawkins, *River out of Eden: A Darwinian View of Life* (New York: Basic Books, 1996), 3.

all things are the products of its intelligence, and living things are just what they always appeared to be—*designed.*" He also explains why there can be cases of "bad design": "If secondary causes, unlike the Primary Cause, are not infallible, if they are defectible, then we might well blame them, rather than the first cause, for any flaws we find (or think we find)."[194]

Thus, even if there are random events during the course of evolution—and there certainly are—this does not mean that evolution follows an erratic or chaotic path. G. K. Chesterton once "seriously joked" about a conspiracy of order in our world of regularity: "One elephant having a trunk was odd, but all elephants having trunks looked like a plot."[195]

The truth of the matter is that natural selection does not replace God and does not take anything away from God. Natural selection selects what is most in accordance with the law and order of God's cosmic design. Cardinal John Henry Newman gave it the right perspective in an 1868 letter: "I do not [see] that 'the accidental evolution of organic beings' is inconsistent with divine design—it is accidental to us, not to God."[196]

The cosmic design creates the "bed" in which the stream of evolution meanders with its successful biological designs. Like water, natural selection follows the path of least resistance against the background of the cosmic design. That's how humanity could arise at the last minute—at 11:59 p.m. on the scale of a day.

---

[194] Augros, *Who Designed the Designer?*, 103.

[195] *The Essential Works of Gilbert K. Chesterton*, vol. 1: Non-Fiction (Radford, VA: Wilder, 2008), 39.

[196] John Henry Newman, "Letter to J. Walker of Scarborough, May 22, 1868," in *The Letters and Diaries of John Henry Newman* (Oxford, UK: Clarendon Press, 1973).

A Final Word

After all we have seen so far, the question of whether God made the earth our home may still be lingering in your mind. Science has acquired so much authority that many people believe that we cannot accept what science cannot prove. This belief gives science arguably too much power, for there are so many questions that science cannot answer, and there is so much that science cannot prove, as we found out. So, is there a final word? If there is, it may not come from science but from religion.

Is God at the Mercy of His Cosmic Design?

Does all that we said about physical constants, laws of nature, and cosmic design mean that God ends up being at the mercy of His own decrees? Does this mean that God Himself has to obey the law and order of the universe He created? Certainly not. If so, divine providence would be a farce. As we said earlier, however, God is the Primary Cause of the universe, not one of its secondary causes. So how can we explain that God is still in charge of the universe in spite of the seemingly "rigid" setting of physical constants, laws of nature, and secondary causes?

Does what we have said so far in this book imply that God can never intervene directly in the natural order? Religion tells us that He does. Theologians distinguish between "mediate" and "immediate" divine providence. The former is exercised through natural secondary causes and can easily be explained without any conflict with science, as we discussed earlier. Stephen Barr uses a helpful analogy that compares God to the author of a play: "The playwright is the cause of the entire play in all its aspects — he pens its every character, event, and word.... We see, then, how idle it is to ask whether some species of beetle exists because it evolved or because God created it."[197] Thinking about God and His creation in this way keeps science and religion out of trouble with each other.

But that might not be the case when it comes to *immediate* divine providence, which does not work indirectly — through natural, secondary causes, that is — but acts directly. This is the case when God heals miraculously, helps us make the right decisions in life, or directly infuses a human soul into a body to make it human. In cases like these, religion claims that the Primary Cause can *directly* steer the chain of secondary causes. It is at this point that scientists would probably protest, as such actions of the Creator seem to violate the laws of nature and have a direct impact on the chains of secondary causes.

Nevertheless, the Church maintains her position regarding immediate divine providence. The *Catechism* confirms this: "The witness of Scripture is unanimous that the solicitude of divine providence is concrete and immediate; God cares for all, from the least things to the great events of the world and its history" (303). In other words, the God of Christianity is not the god of deism. The god of *deism* made the world as a watchmaker who made a watch and let it run its own course, abandoning

---

[197] Barr, *Modern Physics and Ancient Faith*, 262.

it to itself—the hands-off approach, so to speak, of an absent landlord. That is a kind of god that Catholic religion can easily dispense with.

In contrast to deism, *theism* tells us that the Creator of this world remains actively involved with this world, not only by sustaining and preserving what He has created—as a Primary Cause, that is—but also by guiding its course and history directly. The late Cardinal Avery Dulles explicitly stresses this point:

> God continues to act in history. In the course of centuries, he gave revelations to his prophets; he worked miracles; he sent his own Son to become a man; he raised Jesus from the dead. If God is so active in the supernatural order, producing effects that are publicly observable, it is difficult to rule out on principle all interventions in the process of evolution. Why should God be capable of creating the world from nothing but incapable of acting within the world he has made?[198]

Pope Benedict XVI claimed something similar when he rejected the modern notion that God is "allowed" to act in the spiritual domain but not in the material: "God is God, and he does not operate merely on the level of ideas.... If God does not also have the power over matter, then he simply is not God."[199]

In other words, the question here is how God can be acting within the world *directly*, without becoming a factor in the chain of secondary causes and without violating His own laws of nature. To solve this seeming enigma, it might be helpful to start from an often overlooked statement by St. Thomas Aquinas: "God is

---

[198] Avery Dulles, "God and Evolution," *First Things* (October 2007).
[199] Joseph Ratzinger, *Jesus of Nazareth: The Infancy Narratives* (New York: Image Books, 2012), 56–57.

[related] to the universe as the soul [*anima*] is to the body."[200] Notice that Thomas did not say, "God is the soul, and the universe is the body"—for that would be pantheism. In his statement, Thomas used the *analogy* of a person—that is, the relationship between soul and body—to portray the relationship between God and the universe. Elsewhere, Aquinas says, "The king is in the kingdom what the soul is in the body, and what God is in the world."[201] These analogies should be taken only as aids to enhance our understanding of God and are therefore inadequate. Only with that restriction may the relationship between mind and body (or soul and body) help us to understand better the relationship between God and His creation.

Now, to explain how God can still be in charge of His creation in spite of its seemingly rigid setting of physical constants, laws of nature, and secondary causes, it might help us to reflect a bit on our own doings first. We are constantly performing active, steering roles in the world. Think of the numerous "creations" we have produced: cars, planes, computers, surgery, antibiotics—and the list keeps going and growing. In making these "creations," we obey the laws of nature, and at the same time we use them. We are under *their* control and they are under *our* control. We are their servants and their masters at the same time.

What we have to realize, though, is that in doing so, we cannot and do not change the laws of nature, but we do have the ability to use them so as to achieve our purposes and intentions. Our minds have the capacity to employ the laws of nature in such a way that we can achieve the goals we have in mind. It is those purposes that can become causes of actions on their own, so they are definitely possible, even in a rigid world of law and order.

---

[200] *Lectura romana in primum Sententiarum Petri Lombardi*, 2, 17, 1, 1.
[201] Thomas Aquinas, *De Regno*, 1, 12–14.

A Final Word

We all know that our minds have the capacity to steer the laws of nature for the goals we have in mind. The laws of nature cannot be violated, but they can be steered: we can't change the law of gravity, for instance, but we can prevent things from falling by using the proper constraints. Think of certain games. When watching a game on the golf course or on the pool table, we see balls following precisely determined courses of cause and effect; they follow rigid, physical laws. When the players of these games hit a ball, they apply the laws of physics—that is, they use a specific force at a certain angle with a specific impact, leading to a cascade of physical causes and effects.

Yet there is one element that does not fit into this predetermined picture, into this cascade of causes and effects: the players of the games. Although there is a cascade of physical causes and effects in each of these games, there is much more going on in each process, because the players of the game have specific intentions in mind, which elude and transcend the laws of science. Do they go *against* the laws of nature? No, they do not, but they do go freely *beyond* those laws, using them for specific purposes.

People who are unable to look beyond these physical laws and causes are completely missing out on what the game is all about. In their freedom, the players fall outside the realm of physics; they themselves steer the course of the laws of nature from outside the system. They steer the laws of nature in directions of their own choosing.

In other words, my mind is the "soul" of everything I do; it is behind everything my body does though neurons and muscles. In that sense, I might say that the soul is part of everything I do with my body, and yet it is not a physical, material part of my body.

Now the point is this: if we are able to steer the laws of nature, why would God not be able to do the same—in an analogous way, of course? Perhaps the analogy of the working of the human mind can help us better understand the working of God's mind in this world. He is actively present in this world, not by going against the laws of nature and its secondary causes, or by supplementing or replacing them, but rather by letting them be the way they are and yet steering them in a certain direction without overstepping the autonomy of secondary causes. Of course, that is not all there is to it, but perhaps this analogy opens the door for us to understand better what God's presence is like in this universe.

In other words, God does not violate the laws of nature but he can manipulate them for his own purposes. That's what we called Divine Providence before. Just as builders are not a physical part of their buildings yet are an active part of every part of them, so God is not a part of what he created, yet he is actively involved with each and every part of it. God is the First Cause who operates in and through secondary causes.

The analogy of the relationship between mind and body may help us to see how God is the soul and spirit that pervades all creation, and thus is part of everything in Creation without becoming a physical part of it. God is not one of the players on the world scene, but he is the Author and Director of this cosmic play. That's how God works miracles in our world.

On the one hand, this analogy prevents us from thinking that God's activity is a physical, inner-worldly factor interfering in the physical process of evolution, as ID theorists take it. On the other hand, God can still be directly involved with the process of evolution and the course of human history as science tries to decipher it—but without becoming a physical part of it.

In other words, God is not one of the players on the world scene, but He is the Author and Director of this cosmic play. God is not a physical element in the history of the universe, but He is the Lord of history, who keeps everything that's happening in the universe in concert. When we focus only on the laws of nature, we miss out on the beautiful play that is going on around us, when God steers the laws of nature toward a specific outcome. Since He is the source and soul of everything, He is part of each and every cosmic event that's taking place, but without becoming a cosmic part.

A Purpose-Driven World

So far, we have been using expressions such as *divine plan, cosmic design,* and *divine providence.* All of these are somehow connected with the term *purpose.* Each of them suggests that we are living in what some people like to call a "purpose-driven" world.

In the meantime, we should realize that there is a big difference between what God wills and what He allows. God doesn't will earthquakes and mass extinctions, but He allows them when they are a consequence of the laws of nature — as God doesn't will wars but allows them when humans use their freedom to start them.

The term *purpose,* however, has become very controversial in our modern vocabulary. Why? There are probably many reasons, but an important one is the fear of teleological language. Science is partly to blame for this fear. Scientists have become increasingly reluctant to speak in terms of purposes. They like to stress that the daily sunrise is not a purpose of the sun. Planets orbit the sun, but that's not done on purpose. Water seeks its own level, but not purposely. Plants do not intentionally seek the light, but they do respond to the light in their environment through a light-sensing hormone, called auxin, which makes plant shoots curve toward light. Eye patterns on butterfly wings have the effect of warning enemies; that is a function of eye patterns, not a purpose of butterflies. Medications in the blood don't purposely go after their target cells; instead, they combine with their targets like keys fit into locks. In all such cases, purposes have been replaced with causal mechanisms.

In the Beginning

So, in a sense, there is potential trouble looming when we introduce the word *purpose* into the discussion here. It should not surprise us, then, that scientists have banished that word from their vocabulary. As a matter of fact, the concept of purpose was taken out of astronomy by Nicolas Copernicus, out of physics by Isaac Newton, and out of biology by Charles Darwin. This may explain why some scientists obsessively reject any use of that word.

This censure has had quite some consequences. Our world has become a world without purpose, without design, without meaning—certainly not purpose driven. Many scientists like to broadcast this message through media and academia. The chemist Peter Atkins tells us, "Gone is purpose; all that is left is direction. This is the bleakness we have to accept as we peer deeply and dispassionately into the heart of the universe."[202] Or take this one from the biologist Douglas Futuyma: "We need not invoke, nor can we find any evidence for any design, goal, or purpose anywhere in the natural world."[203] And the biologist Richard Dawkins is happy to join this crowd: "The Universe we observe has precisely the properties we should expect if there is, at bottom, no design, no purpose, no evil and no good, nothing but blind pitiless indifference."[204] But perhaps the most arrogant claim comes from the late paleontologist George Gaylord Simpson, who ventured to proclaim from his quasi-scientific pulpit that "man is the result of a purposeless and natural process that did not have him in mind."[205]

---

[202] Peter Atkins, *The Second Law* (New York: Scientific American Library, 1984), 200. In the last paragraph of chapter 9, he seems to express himself more carefully.

[203] Douglas Futuyma, *Evolution* (Sunderland, MA: Sinauer Associates, 2005), 12.

[204] Dawkins, *River out of Eden*, 133.

[205] George G. Simpson, *The Meaning of Evolution* (New Haven: Yale University Press, 1967), 345.

A Final Word

But the question remains: Did science really eliminate purposes? Is this the end of "purpose talk"? Not really. When scientists removed "purposes" from scientific discourse, they removed them as secondary causes in the realm of science, but they left their reference to the First Cause untouched. So, they did not make purposes disappear entirely; they just moved them from inside to outside the scientific domain.

But after doing so, ironically, they commit a striking inconsistency. The scientists mentioned in the previous paragraph cannot make any of their statements as a scientist. First, they eliminate the concept of purpose from science, but then they keep using it by rejecting it. Once "purpose" has been eliminated from science, it can no longer be used, let alone be explained, by science, as it is forever beyond science's reach. Neo-Darwinism, for instance, just does not know whether evolution has a purpose or not, for the simple reason that the word *purpose* does not exist in its vocabulary.

The fact that purposes are missing on scientific maps does not mean that they do not exist at all; they are not completely out of the picture, although they are out of the scientific picture. The fact, for instance, that houses are missing on highway maps, or that human beings are missing on astronomical maps, does not entitle us to deny their existence; whatever we neglect we can never just reject for that reason. Removing purposes from the territory of science may have been a very legitimate move, but it does not entitle us to remove them from our discourse entirely. There are strong indications that they cannot possibly be removed that way at all. Although purpose was taken out of science, science doesn't necessarily have the last word. The final word may very well come from somewhere else: God.

Even though many modern scientists are self-declared agnostics or atheists, a growing number of them are beginning to see a dimension of religious faith behind science. One of them is the late nuclear physicist and Nobel laureate Werner Heisenberg, who said: "The first drink from the cup of natural science makes you atheistic.... But at the bottom of the cup, God is waiting."[206] Pope Pius XII said in 1951 that "true science discovers God in an ever-increasing degree—as though God were waiting behind every door opened by science."[207] And then there is Max Planck, who revolutionized physics with his quantum theory. It was his observation that "the greatest naturalists of all times, men like Kepler, Newton, Leibniz, were inspired by profound religiosity."[208] And then he goes on, "For the believer, God is the *beginning*, for the scientist He is the *end* of all reflections."[209] St. Thomas Aquinas said something similar much earlier: "All our knowledge has its origin in sensation. But God is most remote from sensation. So he is not known to us first, but last."[210] Elsewhere Planck says:

All matter originates and exists only by virtue of a force which brings the particles of the atom to vibration. I must assume behind this force the existence of a conscious and intelligent mind. This mind is the matrix of all matter.[211]

The physicist Paul Davies once summarized this well:

People take it for granted that the physical world is both ordered and intelligible.... Nobody asks where they came

---

[206] Quoted in Ulrich Hildebrand, "Das Universum: Hinweis auf Gott?" *Ethos* 10 (October 1988).

[207] Address to the Pontifical Academy of Sciences, November 22, 1951.

[208] Max Planck, *Religion und Example* (Leipzig: Johann Ambrosius Barth Verlag, 1937), 332.

[209] Max Planck, "Religion and Natural Science (Lecture Given 1937)," *Scientific Autobiography and Other Papers*, trans. F. Gaynor (New York, 1949), 184.

[210] Thomas Aquinas, *Disputed Questions on Truth*, 1, 11.

[211] Max Planck, "*Das Wesen der Materie* [The Nature of Matter]," speech given in Florence, Italy, 1944.

from; at least they do not do so in polite company. However, even the most atheistic scientist accepts as an act of faith that the universe is not absurd, that there is a rational basis to physical existence manifested as law-like order in nature that is at least partly comprehensible to us.[212]

So do we want to bring purposes back into the discussion? Part of the answer is no. Purposes do not belong in science because they are not measurable, quantifiable, material, or physical. But those who agree with this should then stick with their decision to discard purposes from science and accept its consequences. Once you remove purposes from scientific discourse, you cannot keep referring to them, at least not in science.

On the other hand, part of the answer is also a definitive yes. If there were no purpose in the universe at all, how would we ever know there is no such thing as a purpose? As C. S. Lewis put it, "If there were no light in the universe and therefore no creatures with eyes, we would never know it was dark."[213]

Besides, I would like to ask those who deny the existence of purposes what the purpose is of trying to prove that there is no purpose in life. As a matter of fact, denying that there are purposes in life defeats its own claim. If it is someone's purpose to remove all purposes from life, he is also wiping out his own purpose of doing so. Hence, those whose purpose it is to eradicate all purposes from life have lost even the very purpose for doing so.

I don't think any sane person would deny that humans have purposes in life. The human mind enables us to articulate purposes, and free will lets us work hard so they may come true. Since we are made in the image and likeness of God, the Divine Mind must

---

[212] Paul Davies, "Physics and the Mind of God: The Templeton Prize Address," *First Things* (August 1995).

[213] C. S. Lewis, *Mere Christianity* (San Francisco: Harper, 2001), 46.

have purposes too. But as the book of Proverbs tells us, "Many are the plans in the mind of a man, but it is the purpose of the Lord that will be established" (19:21, RSVCE). That's what we call divine providence—all the plans God has for each of us and for our universe. That's where we find, according to the *Catechism*, "the definitive, superabundant answer to the questions that man asks himself about the meaning and purpose of his life" (68).

If this analysis is right, we are still allowed to talk about the universe in terms of purposes, as long as we don't do so with scientific pretentions. In other words, there is indeed a purpose in the physical constants in the universe. There is indeed a purpose in the laws of nature. There is indeed a purpose in the largeness of the universe. There is indeed a purpose in the age of the universe. There is indeed a purpose in the way Planet Earth is configured. There is indeed a purpose in the path biological evolution has followed.

All of this has taken place in the "bed" of the cosmic design, steered by a divine plan based on divine providence. The entire process that our universe has gone through had to be the way it was, simply because it was intended this way. In other words, we were meant to be here. The earth was made for us. Everything was indeed fine-tuned for our coming. We live in a purpose-driven universe. Gone are chaos, mere randomness, and utter purposelessness.

Let's quote two important sources again. The *Catechism* declares emphatically that the world "is not the product of any necessity whatever, nor of blind fate or chance" (295). And Pope Benedict XVI insisted: "We are not some casual and meaningless product of evolution. Each of us is the result of a thought of God. Each of us is willed, each of us is loved, each of us is necessary."[214]

---

[214] Benedict XVI, Homily at the Mass for the inauguration of his pontificate (April 24, 2005).

A Final Word

We need to ask the question again: Did God make the earth our home? Hugh Ross nicely summarizes our conclusions about Planet Earth with these words: "The ideal place for any kind of life as we know it turns out to be a solar system like ours, with a galaxy like the Milky Way, within a supercluster of galaxies like the Virgo supercluster, within a super-supercluster like the Laniakea super-supercluster."[215]

[215] Ross, *Improbable Planet*, 28.

Appendix

Is Planet Earth Unique?

For generations, mankind has gazed upon the heavens and wondered, "Are we alone in the universe?" Science-fiction writers such as Isaac Asimov and Herbert George (H.G.) Wells brought their fanciful ideas about extraterrestrial life to the general public in the last century. Many popular books and films either revolve around or are influenced by the notion of extraterrestrial life. Ironically, it is the very inhabitants of Planet Earth who create this kind of science fiction and are able to enjoy it. We might be the only ones in the universe who can even think about that possibility.

The fact remains, though, that Planet Earth has become our home. But that doesn't mean it's the only place in the universe that could be our home. We must be honest about the fact that we don't really know whether it is or not. Earth is unquestionably an *exceptional* place in the universe, but perhaps not a *unique* place.[216]

Yet some people believe there are other places, perhaps even many, where life, even intelligent life, might exist. Their reason for thinking so is usually based on a follow-the-water strategy and the statistical law of large numbers. If there is solid rock and water

---

[216] Ross, *Improbable Planet*, 80.

on any planet among the numerous planets in our universe, then that planet could be our home as well.

As we said earlier, the cosmic design of the universe comes with the potentiality for life. But that doesn't necessarily mean that life is ubiquitous in the universe—more than that is needed. Why not life on the moon, for instance? Well, life needs much more than just liquid water. It also needs supportive nutrients in the form of minerals, most of which depend on plate tectonics and a strong magnetic field. To manifest such features, a moon's mass must be at least 23 percent of Earth's mass. For these features to last more than a few billion years requires a mass and density closely equivalent to Earth's.[217] To get a moon as massive as Earth would require, among some other challenges, a host planet at least twenty times more massive than Jupiter. Such a massive planet, however, would inevitably generate orbital chaos for other planets in its system.[218]

In other words, the follow-the-water strategy and the law of large numbers are not enough to find life somewhere else in the universe. True, after hydrogen, water is the most abundant molecule in the universe. But what about radiation? For life's sake, too much or too little radiation is dangerous.[219] In addition, stars more massive than the sun exhibit more extreme variation in ultraviolet emission, as do stars less massive than the sun.[220] And what about atmospheric pressure? When pressure is low, such as on Mars, a drop of water evaporates in a second. And what about so many other conditions we find on Planet Earth, which enabled

---

[217] Vlada Stamenkovic, Lena Noack, Doris Breuer, and Tilman Spohn, "The Influence of Pressure-Dependent Viscosity on the Thermal Evolution of Super-Earths," *Astrophysical Journal* 748 (March 2012): 41.

[218] Ross, *Improbable Planet*, 231–233.

[219] Ibid., 40–41.

[220] Ibid., 87.

life and humanity to emerge here? A planet closer than 95 percent of Earth's distance from the sun would experience runaway evaporation, whereas a planet beyond 137 percent of that distance would experience runaway freezing.[221] And when a planet orbits too close to its star, it becomes tidally locked in the same position (as the moon is tidally locked with Earth). In such a case, if there were life there, it could exist only in the twilight zone between permanent light and permanent darkness.

It's true that, in spite of all these conditions and considerations, there still might be life on other planets in the universe. More than three thousand exoplanets—planets that, like ours, are close to a sun—have been identified by NASA's Kepler space probe mission.[222] Would they qualify for life, even intelligent life? Only time can tell. All we can say at this point is that the history of Planet Earth gives us the impression that our planet is highly improbable, exceptional, and perhaps even utterly unique. But that's not an unbeatable argument against the possibility of life on other planets. The best it can do is make us more skeptical about the follow-the-water strategy.

Opinions about the issue of extraterrestrial life vary greatly along a broad spectrum. At one end of the scale is the camp of those who claim there *cannot* be life on other planets. At the other end are those who claim that there *must* be life on other planets. Probably the majority of scientists belong to the latter group. Some of them have become obsessed with the idea of extraterrestrial life. There might be many explanations for their fascination. One of them is their belief in the statistical law of large numbers: given the nearly

[221] James F. Kasting, Daniel P. Whitmire, and Ray T. Reynolds, "Habitable Zones around Main Sequence Stars," *Icarus* 101 (January 1993): 108–128.

[222] See "How Many Exoplanets Has Kepler Discovered?" Kepler and K2, last updated October 25, 2018, https://www.nasa.gov/kepler/discoveries.

infinite number of planets, there must be a few of them with life on them. Another one is their almost instinctive reaction against what had falsely been believed for centuries—that we live in a geocentric and anthropocentric universe. And then there is their claim that extraterrestrial life *must* exist, because their argument seems so compelling: if some of those planets have water, and if some of these planets have atmospheres like Earth's, and if some of those Earth-like planets spontaneously generate amino acids, and if some of those amino acids result in higher life-forms, then life is bound to exist on other planets. Well, that's a lot of *ifs*.

No matter what, the search for life on other planets is on. But science doesn't have a good track record when it comes to claims about extraterrestrial life, let alone intelligent life on other planets. Not too long ago, Percival Lowell, founder of the Lowell Observatory, believed, on the basis of what he called "careful scientific observation," that there were Martian-made canals on Mars.[223] It is hard to tell where science changes into science fiction. NASA is a major player in the concerted effort to search for extraterrestrial life. From its beginning, NASA speculated that evidence of past intelligent life may lie somewhere in our solar system. In 1996, NASA's Johnson Center cautiously announced that they had discovered fossilized microbes in meteorite ALH84001, known to have come from Mars. Under an electron microscope, they found what appeared to be small, worm-like objects on parts of the rock. This discovery is still being debated, but many argue that this was simply an unintended contamination.[224] And there have been similar extravagant claims since.

[223] Robert Crossley, "Percival Lowell and the History of Mars," *Massachusetts Review* 41, no. 3 (Autumn 2000): 297–318.

[224] David McNab and James Younger, *The Planets* (London: Yale University Press, 1999), 213.

Very recently, a video on YouTube suggested that the space agency was about to announce the discovery of life beyond Earth. The video made such a big splash online that NASA science chief Thomas Zurbuchen had to address the rumor with a tweet saying, "Contrary to some reports, there's no pending announcement from NASA regarding extraterrestrial life."[225] The borderline between science and science fiction had become blurry again—perhaps too blurry.

Obviously, the question of whether life exists on other planets is a scientific issue that can never be settled ahead of time. But we should address one more, important issue in this context. The late Supreme Court judge Antonin Scalia told the story of the best lesson he learned during his studies at Georgetown University.[226] It happened during his oral comprehensive examination at the end of his senior year. His history professor, Dr. Wilkinson, asked him one last question: "Of all the historical events you have studied, which one in your opinion had the most impact upon the world?" Scalia mulled over several options: the French Revolution, the Battle of Lepanto, or perhaps the American Revolution? Whatever he eventually happened to answer, Dr. Wilkinson informed him of the right answer. Of course, it was the Incarnation—the moment God came to Earth in the person of Jesus Christ. That he considered the one real turning point in the history of Planet Earth!

The Incarnation is a part of Earth's history that we cannot just ignore when we talk about the possibility or probability of intelligent life on other planets. If we ever discover that there is life, especially intelligent life, on other planets, such a discovery might have quite some theological repercussions. Was the rest of

---

[225] Thomas Zurbuchen (@Dr_ThomasZ), Twitter, June 26, 2017, 9:16 a.m., https://twitter.com/dr_thomasz/status/87937283789471744 1?lang=en.

[226] Antonin Scalia, *Scalia Speaks: Reflections on Law, Faith, and Life Well Lived* (New York: Crown Forum, 2017), 147.

the universe also affected by the Incarnation in Bethlehem? Or do we have to come up with other answers?

The fact remains that Planet Earth is highly improbable, perhaps even unique. This could mean that we may never have to deal with such theological questions. But if we ever will, it is interesting to know that Catholics thought about such questions long before NASA came along. Some have said that the discovery of extraterrestrial life could perhaps remind us of the awesome creativity of our God. God's imagination, after all, cannot be constrained by the limits of our human imagination and by our quest for knowledge. For example, the French priest and philosopher John Buridan (1295–1363) thought that denying the possibility of life on other worlds implied imposing a limit to God's power. He wrote, "We hold from faith that just as God made this world, so he could make another or several worlds."[227] And Cardinal Nicholas of Cusa (1401–1464) went even further when he wrote, "We surmise that none of the other regions of stars are devoid of inhabitants."[228]

Voices like these can still be heard in our time. Jesuit Brother Guy Consolmagno, a planetary scientist of the Vatican Observatory, thinks that the prospect of intelligent life elsewhere in the universe could be theologically fruitful. In a 2002 interview with *U.S. Catholic*, he said about extraterrestrial intelligent beings, "I think we recognize that if they're from Alpha Centauri or from the next galaxy over, they're still God's creation. It's all God's creation! If we ever find intelligent life, we'll have an interesting dialogue about the nature of the Incarnation." Consolmagno has

---

[227] Cited in S. J. Dick, ed., *Many Worlds: The New Universe, Extra-Terrestrial Life and the Theological Implications* (Philadelphia: Templeton Foundation Press, 2000), 29.

[228] Nicholas of Cusa, *De Docta Ignorantia*, 2, 12, 171.

also written a book titled *Would You Baptize an Extraterrestrial?* Although Br. Guy does not believe that we have any conclusive proof of extraterrestrials as of yet, he responds to the question by saying, "Yes. If she asks for it."[229] Pope Francis, well known for his off-the-cuff remarks, said during morning Mass in May 2014, "If an expedition of Martians arrives and some of them come to us and if one of them says: 'Me, I want to be baptized!' what would happen?" At least the pontiff left the possibility open, without giving a definite answer. The answer "Yes, if they ask for it" is probably the best answer for now.

Where does this leave Catholics? According to a 2011 study for the Royal Society, about 90 percent of believers felt that if intelligent life were to be discovered on other planets, they would not have a crisis of faith.[230] According to a 2015 study by Joshua Ambrosius, professor at the University of Dayton, Catholics and people with no religious belief are the two groups most optimistic about the possibility of discovering extraterrestrial life in the next forty years.[231] The new fad seems to be that we have "brothers and sisters" on other planets. But of course, we need answers from science rather than from polls.

On the other hand, the inconsistencies and failures of scientism should warn us that science does not necessarily have the final word. Perhaps we can learn a few lessons from what scientists have said

---

[229] Guy Consolmagno, S.J., and Paul Mueller, S.J., *Would You Baptize an Extraterrestrial? . . . And Other Questions from the Astronomers' In-Box at the Vatican Observatory* (New York: Doubleday Image Books, 2014).

[230] Martin Dominik and John C. Zarnecki, "The Detection of Extra-Terrestrial Life and the Consequences for Science and Society," *Philosophical Transactions of the Royal Society* (January 2011).

[231] Presented at a November 2014 conference of the Society for the Scientific Study of Religion in Indianapolis.

in the past about the universe. Judeo-Christian religion has always asserted that the universe had a beginning in time. Not only do the first words of Genesis "In the beginning," suggest this view, but also the Fourth Lateran Council declared in 1215 that the world was created by God "from the beginning of time" (*ab initio temporis*). Interestingly, most scientists around the turn of the twentieth century were of the opinion that the universe could not have a beginning. As mentioned earlier, before scientists embraced the notion that the universe was created as the result of the Big Bang, it was commonly believed that the size of the universe was an unchanging constant. As late as 1959, a survey of leading American astronomers and physicists found that two-thirds of them believed that the universe had no beginning. Nowadays, there is hardly any scientist who still believes that. The Big Bang theory now has the majority vote. In this case, it would not have been wise for the Catholic Church to bend her teaching to prevailing scientific opinions of the past.

Perhaps something similar holds for the issue of extraterrestrial life, including intelligent life. Religion should not bend too easily for science, and certainly not for science fiction. As we discovered in this book, there are strong indications that Planet Earth is highly improbable and exceptional, and perhaps even unique. This becomes even truer when we add the religious doctrine of the Incarnation to the special features of Planet Earth. Instead of saying there *must* be intelligent life on other planets, we should say that, at best, there *may* be intelligent life on other planets. Let's not jump to conclusions yet. But no matter what, let's not forget that Jesus' Church is "catholic," which is the Greek word for "about the whole." She is global (for the entire globe) as well as universal (for the entire universe).

When people raise the age-old question "Are we alone in the universe?" they can receive only speculative and hypothetical answers until we have more evidence pro or con. So far, science has not been able to tell us that we are not alone in the universe.

From a logical point of view, it is much harder to prove that we are alone—which is the assertion of a "universal negative"—than that we are *not* alone. For instance, it is much harder to establish that there are *no* black swans in the universe than to prove that there is at least one somewhere on a planet. So, we are waiting for positive evidence. All we can say right now is that science has not been able to give us that positive evidence. The best we can say at this moment is that Planet Earth is a highly improbable phenomenon, and therefore perhaps even utterly unique—a place specifically made for us, a place where we were meant to be.

In essence, though, it doesn't really matter whether we are alone in the universe or not. Christianity tells us we are never alone. As the psalmist puts it, "If I take the wings of dawn and dwell beyond the sea, even there your hand guides me, your right hand holds me fast." (139:9–10). And the prophet Isaiah says, "Can a mother forget her infant, be without tenderness for the child of her womb? Even should she forget, I will never forget you. See, upon the palms of my hands I have engraved you" (49:15–16). It is at the core of the Judeo-Christian message that God holds each of us in the palm of His hand—no matter how tiny we are and how tiny our planet is in the whole universe. In that sense, the message of the Book of Nature is not too far off from the message of the Book of Scripture.

Psalm 8 summarizes all of this well in a timeless manner:

When I see your heavens, the work of your fingers,
    the moon and stars that you set in place—
    What is man that you are mindful of him,
    and a son of man that you care for him?
Yet you have made him little less than a god,
    crowned him with glory and honor.
You have given him rule over the works of your hands,
    put all things at his feet. (vv. 4–7)

Index

Ambrosius, Joshua, 173
annihilation, 29
anthropic coincidences, 44
anthropic principle, 48
Aquinas, 24, 34, 37, 56, 64, 89, 106, 130, 155, 162
Archimedes, 39
Atkins, Peter, 26, 71, 134, 160
atmosphere, 80, 118
atomic number, 13
Augros, Michael, 51, 150
Augustine, 4, 22, 126

Baronius, Cardinal, 4
Barrow, John D., 55
Barr, Stephen, 23, 28, 45, 52, 54, 63, 75, 76, 77, 130, 148, 154
Beckwith, Francis J., 107
Benedict XVI, 4, 36, 78, 155, 164
Big Bang, 8
Big Bang theory, 19
Bostrom, Nick, 48
Buridan, John, 172

Carroll, William E., 25, 29, 89
Casimir, Hendrik, 144
Cassirer, Ernst, 123
chemosynthesis, 95
Chesterton, G. K., 142, 151

coincidental, 73
Collins, Francis, 94, 102
concept, 122
Consolmagno, Guy, 172
continental drift, 86
contingent, 130
Copernicus, Nicolaus, 160
cosmological constant, 47
Creation, 21, 24
Creation out of nothing, 24
Crick, Francis, 135, 136
Cusa, Nicholas of, 172
cyanobacteria, 97

Darwin, Charles, 146, 160, 181
Davies, Paul C., 55, 61, 64, 162
Dawkins, Richard, 134, 150, 160
death, 128
de Duve, Christian, 110
deep-sea vent, 95
deism, 154
deuterium, 70
Divine Providence, 76, 154, 158
Dobzhansky, Theodosius, 128
Dulles, Avery, 33, 155
Dyson, Freeman, 53

Eddington, Arthur, 102

Einstein, Albert, 21, 47, 54, 59, 60, 137, 140
Eldredge, Niles, 94
electromagnetic fine structure constant, 43
electromagnetic force, 46
extraterrestrial life, 167

Feser, Edward, 35, 140
Feyerabend, Paul, 139
fine structure constant, 46
free radicals, 109
Freud, Sigmund, 33
Futuyama, Douglas, 160

galaxies, 65
Galileo, 4
general relativity, 47
Genesis 1, 5, 22
grand unified theory, 40, 142
gravity, 26, 45, 70
Great Oxygenation Event, 98

half-life, 14, 74
Hart, David Bentley, 23
Hawking, Stephen, 25, 48, 51, 53, 55, 138, 144
Heisenberg, Werner, 161
Hoyle, Fred, 19, 54
Hubble constant, 11
Hubble, Edwin, 19

ice age, 115
intelligent design theory, 102
intelligibility, 59, 142
inverse square law, 40, 45
isotopes, 13, 82, 93
isotopic signature, 87

Jaki, Stanley, 107
Jeans, James, 62
John Paul II, 61

Kass, Leon, 149
Kepler, Johannes, 61
Kreeft, Peter, 32

language, 123
law of gravity, 39
law of large numbers, 9, 41, 50, 51, 72, 74, 101, 167, 169
laws of nature, 36, 39
Leibniz, Gottfried, 107
Lemaître, Georges, 20, 63
Lewis, C. S., 108, 163
life signature, 93
luminosity, 81

magnetic field, 82
Maslow, Abraham, 140
mass extinction, 111
Mayr, Ernst, 142
Milky Way, 66
Miller-Urey experiment, 92
mind, 127
minerals, 99
monotheism, 7
moon, 86
Murray, Joseph, 61
mutation, 74

natural selection, 147
neuron, 130
Newman, John Henry, 151
Newton, Isaac, 39, 40, 106, 123, 146, 160

order, 59, 142

Paley, William, 45
Pangaea, 86
Pangea, 112
Pascal, Blaise, 57, 73
Periodic Table of Elements, 13
Perry, Ralph Barton, 141

Index

photosynthesis, 95, 97
physical constants, 42, 44
Pius XII, 6, 162
Planck, Max, 44, 138, 162
plate tectonics, 99
Polkinghorne, John, 28
Pope, Alexander, 78
Precambrian, 80
Primary Cause, 35, 107, 151
purpose, 160

quantum tunneling, 28

radioactive dating, 13
radioactive decay, 14, 41, 42, 74
randomness, 72
red-shift, 20
religion, 130
rights, 125
RNA-world hypothesis, 91
Ross, Hugh, 82, 99, 114, 165
Royal Society of London, 143
Ryle, Gilbert, 139

Sagan, Carl, 25, 57
Scalia, Antonin, 171
scientism, 135
self-awareness, 126

Simpson, George Gaylord, 160
Skhul cave, 128
Smolin, Lee, 25
solar system, 67
Spencer, Herbert, 148
spontaneous creation, 26
strong nuclear force, 42, 46
Suárez, Francisco, 107

tectonic plates, 119
teleology, 149
terrestrial planets, 68
theism, 155
Thomas Aquinas. *See* Aquinas
Tipler, Frank J., 55
Townes, Charles, 61

Ussher, James, 3

vacuum, 28
Vilenkin, Alexander, 25
volcanism, 120

Wallace, Alfred, 148
weak nuclear force, 42
Wheeler, John A., 55
Wittgenstein, Ludwig, 122

Zeeman, Pieter, 138

About the Author

Gerard M. Verschuuren is a human biologist, specializing in human genetics. He also holds a doctorate in the philosophy of science. He studied and worked at universities in Europe and in the United States. Currently semiretired, he spends most of his time as a writer, speaker, and consultant on the interface of science and religion, faith, and reason.

Some of his most recent books are: *Darwin's Philosophical Legacy: The Good and the Not-So-Good*; *God and Evolution?: Science Meets Faith*; *The Destiny of the Universe: In Pursuit of the Great Unknown*; *It's All in the Genes!: Really?*; *Life's Journey: A Guide from Conception to Growing Up, Growing Old, and Natural Death*; *Aquinas and Modern Science: A New Synthesis of Faith and Reason*; *Faith and Reason: The Cradle of Truth*; *The Myth of an Anti-Science Church: Galileo, Darwin, Teilhard, Hawking, Dawkins*; and *At the Dawn of Humanity: The First Humans*.

For more information, visit https://en.wikipedia.org/wiki/Gerard_Verschuuren.

Verschuuren can be contacted at www.where-do-we-come-from.com.

Sophia Institute

Sophia Institute is a nonprofit institution that seeks to nurture the spiritual, moral, and cultural life of souls and to spread the Gospel of Christ in conformity with the authentic teachings of the Roman Catholic Church.

Sophia Institute Press fulfills this mission by offering translations, reprints, and new publications that afford readers a rich source of the enduring wisdom of mankind.

Sophia Institute also operates the popular online resource CatholicExchange.com. *Catholic Exchange* provides world news from a Catholic perspective as well as daily devotionals and articles that will help readers to grow in holiness and live a life consistent with the teachings of the Church.

In 2013, Sophia Institute launched Sophia Institute for Teachers to renew and rebuild Catholic culture through service to Catholic education. With the goal of nurturing the spiritual, moral, and cultural life of souls, and an abiding respect for the role and work of teachers, we strive to provide materials and programs that are at once enlightening to the mind and ennobling to the heart; faithful and complete, as well as useful and practical.

Sophia Institute gratefully recognizes the Solidarity Association for preserving and encouraging the growth of our apostolate over the course of many years. Without their generous and timely support, this book would not be in your hands.

www.SophiaInstitute.com
www.CatholicExchange.com
www.SophiaInstituteforTeachers.org

Sophia Institute Press® is a registered trademark of Sophia Institute.
Sophia Institute is a tax-exempt institution as defined by the
Internal Revenue Code, Section 501(c)(3). Tax I.D. 22-2548708.